FIELD BOOK FOR QUALITY CONTROL IN EARTHWORK OPERATIONS

FIELD BOOK FOR QUALITY CONTROL IN EARTHWORK OPERATIONS

Project Construction Management Book

A L B E R T O   M U N G U I A   M I R E L E S

FIELD BOOK FOR QUALITY CONTROL IN EARTHWORK OPERATIONS
PROJECT CONSTRUCTION MANAGEMENT BOOK

iUniverse books may be ordered through booksellers or by contacting:

iUniverse LLC
1663 Liberty Drive
Bloomington, IN 47403
www.iuniverse.com
1-800-Authors (1-800-288-4677)

ISBN: 978-1-4917-4481-9 (sc)
ISBN: 978-1-4917-4483-3 (hc)
ISBN: 978-1-4917-4482-6 (e)

Library of Congress Control Number: 2014915070

Printed in the United States of America.

iUniverse rev. date: 08/27/2014

Field Book for Quality Control in Earthwork Operations

Project Name: _____

City: _____

County: _____

Earthwork Contractor: _____

Inspection Company: _____

Contact Information

Name: _____

Phone No.: _____

Book No. ___ of ___ _____

Contents

A Note To Readers

This field book will help you to properly document in-field moisture-density tests performed during earthwork operations in highways on items such as embankment, subgrade, bases, backfilling trenches for water lines, sewer lines, storm drains, and structural backfill around concrete structures. This field book was created after analyzing the most effective methods in monitoring and controlling the manipulation of dirt by succesful contractors, combined with years of experience dealing with project quality management in my career of providing civil engineering services during the construction and inspection of infrastructure projects.

This field book covers a tested methodology that, if used well and daily, will provide a systematic collection of information to monitor in-field moisture-density results to determine whether a contractor complies with quality requirements and to identify ways to eliminate the causes of unsatisfactory performance.

This is not a normal project-quality management book, intended to be read and then stored on a shelf. *This notebook contains blank sheets to enter the testing done as it happens.* So it is intended to be carried out to the job site on a daily basis.

But it is also a book. *It is a source of information and training about the minimum knowledge required to master the art of dealing with dirt.* This is, then, a *field book* where you record the data in field and have on hand information that might help you solve a common problem dealing with extracting, hauling, or placing dirt during the construction of highways in Texas.

This field book is divided in two sections with logs in which testing data can be recorded, six lessons providing the framework for earthwork operations, and three informative tables that provide practical information.

Please visit my webpage at www.cs4highway.com, and register to receive free training on subjects related to highway construction and to receive updates when more information is available.

Introduction

Soil testing is a quantitative method for controlling compacted fill material, and the actual number and types of test made depend on the requirements specified by the designer. The Texas Department of Transportation established these requirements in a document called *Schedule for Testing and Sampling*. Soil testing is specified there.

The Texas Department of Transportation also defines Quality Assurance and Quality Control as the following:

- *Quality Assurance (QA)*—Sampling, testing, inspection, and other activities conducted by the engineer to determine payment and make acceptance decisions
- *Quality Control (QC)*—Sampling, testing, and other process control activities conducted by the contractor to monitor production and placement operations

In other words, quality assurance audits are performed by the owner's representative, and quality control testing is performed by the contractor. This is because the purpose of performing quality-assurance processes is to ensure that the project satisfies the quality standards as well as stakeholders' expectations while the purpose of the quality-control process is to identify the causes of poor processes or product quality and to recommend a solution

You work for either the owner or the contractor. This field book is helpful because testing has to be done onsite, and now you have a place to record while in the field for further analysis and transmittal of the result.

Consec. #	1	2	3	4	5	6
Date	3-14-12	3-15-12	3-15-12	3-16-12	3-17-12	3-20-12
Test #	21	24	25	30	31	33
Chain	Main lane	South Bound	Frontage Rd	Frontage Road 3	North Bound	North Bound
Sta.	264+30	280+00	85+00	30+35	322+00	322+00
CL Offset	20 Rigth	30 R	10 Left	5 L	20 R	20R
Lift #	1	1	2	2	23	23
Material Type	C	C	C	C	C	C
Proctor #	23	3	27	16	32	32
Da	102.2	108.6	114.0	111.8	112.7	112.7
Optimum Moisture	21.5	17.9	14.9	16.4	13.9	13.9
Plasticy Index	25	21	13	14	8	8
Liquid Limit	48	38	30	36	25	25
Density Req.	98-102	98-102	98	98	98	98
Depth	12"	12"	12"	12"	12"	12"
Moisture Standard.	444	449	449	444	555	552
Density Standard	2137	2160	2160	2143	1822	1814
Density Count	469.4	404.5	398.3	400.3	423.0	302
Moisture Count	217.9	212	172.5	182.0	189.0	204
Wet Density	125.4	131.4	132.2	131.6	124.0	129.4
Dry Density	101.0	108.0	113.6	111.6	105.5	111.2
% Moisture	24.1	21.6	16.3	18	17.5	16.3
% Compaction	98.8	99.5	99.7	99.8	93.6	98.7
Pass / fail	Pass	Pass	Pass	Pass	Fail	Pass
Inspector	A.M	A M	A.M	A.M	A.M	A.M
Lab Tech	R.1	R.1	R.1	R.1	R.1	R.1
Lab report #	11-C-0713	11-C-0714	11-0715	11-C-0716	11-C-0717	11-C-0718 Re-Test.

SECTION 1
DENSITY-MOISTURE DAILY LOGS

Consec. #						
Date						
Test #						
Chain						
Sta.						
CL Offset						
Lift #						
Material Type						
Proctor #						
Da						
Optimum Moisture						
Plasticy Index						
Liquid Limit						
Density Req.						
Depth						
Moisture Standard.						
Density Standard						
Density Count						
Moisture Count						
Wet Density						
Dry Density						
% Moisture						
% Compaction						
Pass / fail						
Inspector						
Lab Tech						
Lab report #						

Consec. #						
Date						
Test #						
Chain						
Sta.						
CL Offset						
Lift #						
Material Type						
Proctor #						
Da						
Optimum Moisture						
Plasticy Index						
Liquid Limit						
Density Req.						
Depth						
Moisture Standard.						
Density Standard						
Density Count						
Moisture Count						
Wet Density						
Dry Density						
% Moisture						
% Compaction						
Pass / fail						
Inspector						
Lab Tech						
Lab report #						

Consec. #						
Date						
Test #						
Chain						
Sta.						
CL Offset						
Lift #						
Material Type						
Proctor #						
Da						
Optimum Moisture						
Plasticy Index						
Liquid Limit						
Density Req.						
Depth						
Moisture Standard.						
Density Standard						
Density Count						
Moisture Count						
Wet Density						
Dry Density						
% Moisture						
% Compaction						
Pass / fail						
Inspector						
Lab Tech						
Lab report #						

Consec. #						
Date						
Test #						
Chain						
Sta.						
CL Offset						
Lift #						
Material Type						
Proctor #						
Da						
Optimum Moisture						
Plasticy Index						
Liquid Limit						
Density Req.						
Depth						
Moisture Standard.						
Density Standard						
Density Count						
Moisture Count						
Wet Density						
Dry Density						
% Moisture						
% Compaction						
Pass / fail						
Inspector						
Lab Tech						
Lab report #						

Consec. #					
Date					
Test #					
Chain					
Sta.					
CL Offset					
Lift #					
Material Type					
Proctor #					
Da					
Optimum Moisture					
Plasticy Index					
Liquid Limit					
Density Req.					
Depth					
Moisture Standard.					
Density Standard					
Density Count					
Moisture Count					
Wet Density					
Dry Density					
% Moisture					
% Compaction					
Pass / fail					
Inspector					
Lab Tech					
Lab report #					

Consec. #						
Date						
Test #						
Chain						
Sta.						
CL Offset						
Lift #						
Material Type						
Proctor #						
Da						
Optimum Moisture						
Plasticy Index						
Liquid Limit						
Density Req.						
Depth						
Moisture Standard.						
Density Standard						
Density Count						
Moisture Count						
Wet Density						
Dry Density						
% Moisture						
% Compaction						
Pass / fail						
Inspector						
Lab Tech						
Lab report #						

Consec. #						
Date						
Test #						
Chain						
Sta.						
CL Offset						
Lift #						
Material Type						
Proctor #						
Da						
Optimum Moisture						
Plasticy Index						
Liquid Limit						
Density Req.						
Depth						
Moisture Standard.						
Density Standard						
Density Count						
Moisture Count						
Wet Density						
Dry Density						
% Moisture						
% Compaction						
Pass / fail						
Inspector						
Lab Tech						
Lab report #						

Consec. #						
Date						
Test #						
Chain						
Sta.						
CL Offset						
Lift #						
Material Type						
Proctor #						
Da						
Optimum Moisture						
Plasticy Index						
Liquid Limit						
Density Req.						
Depth						
Moisture Standard.						
Density Standard						
Density Count						
Moisture Count						
Wet Density						
Dry Density						
% Moisture						
% Compaction						
Pass / fail						
Inspector						
Lab Tech						
Lab report #						

Consec. #						
Date						
Test #						
Chain						
Sta.						
CL Offset						
Lift #						
Material Type						
Proctor #						
Da						
Optimum Moisture						
Plasticy Index						
Liquid Limit						
Density Req.						
Depth						
Moisture Standard.						
Density Standard						
Density Count						
Moisture Count						
Wet Density						
Dry Density						
% Moisture						
% Compaction						
Pass / fail						
Inspector						
Lab Tech						
Lab report #						

Consec. #					
Date					
Test #					
Chain					
Sta.					
CL Offset					
Lift #					
Material Type					
Proctor #					
Da					
Optimum Moisture					
Plasticy Index					
Liquid Limit					
Density Req.					
Depth					
Moisture Standard.					
Density Standard					
Density Count					
Moisture Count					
Wet Density					
Dry Density					
% Moisture					
% Compaction					
Pass / fail					
Inspector					
Lab Tech					
Lab report #					

Consec. #						
Date						
Test #						
Chain						
Sta.						
CL Offset						
Lift #						
Material Type						
Proctor #						
Da						
Optimum Moisture						
Plasticy Index						
Liquid Limit						
Density Req.						
Depth						
Moisture Standard.						
Density Standard						
Density Count						
Moisture Count						
Wet Density						
Dry Density						
% Moisture						
% Compaction						
Pass / fail						
Inspector						
Lab Tech						
Lab report #						

Consec. #						
Date						
Test #						
Chain						
Sta.						
CL Offset						
Lift #						
Material Type						
Proctor #						
Da						
Optimum Moisture						
Plasticy Index						
Liquid Limit						
Density Req.						
Depth						
Moisture Standard.						
Density Standard						
Density Count						
Moisture Count						
Wet Density						
Dry Density						
% Moisture						
% Compaction						
Pass / fail						
Inspector						
Lab Tech						
Lab report #						

Consec. #						
Date						
Test #						
Chain						
Sta.						
CL Offset						
Lift #						
Material Type						
Proctor #						
Da						
Optimum Moisture						
Plasticy Index						
Liquid Limit						
Density Req.						
Depth						
Moisture Standard.						
Density Standard						
Density Count						
Moisture Count						
Wet Density						
Dry Density						
% Moisture						
% Compaction						
Pass / fail						
Inspector						
Lab Tech						
Lab report #						

Consec. #					
Date					
Test #					
Chain					
Sta.					
CL Offset					
Lift #					
Material Type					
Proctor #					
Da					
Optimum Moisture					
Plasticy Index					
Liquid Limit					
Density Req.					
Depth					
Moisture Standard.					
Density Standard					
Density Count					
Moisture Count					
Wet Density					
Dry Density					
% Moisture					
% Compaction					
Pass / fail					
Inspector					
Lab Tech					
Lab report #					

Consec. #					
Date					
Test #					
Chain					
Sta.					
CL Offset					
Lift #					
Material Type					
Proctor #					
Da					
Optimum Moisture					
Plasticy Index					
Liquid Limit					
Density Req.					
Depth					
Moisture Standard.					
Density Standard					
Density Count					
Moisture Count					
Wet Density					
Dry Density					
% Moisture					
% Compaction					
Pass / fail					
Inspector					
Lab Tech					
Lab report #					

Consec. #					
Date					
Test #					
Chain					
Sta.					
CL Offset					
Lift #					
Material Type					
Proctor #					
Da					
Optimum Moisture					
Plasticy Index					
Liquid Limit					
Density Req.					
Depth					
Moisture Standard.					
Density Standard					
Density Count					
Moisture Count					
Wet Density					
Dry Density					
% Moisture					
% Compaction					
Pass / fail					
Inspector					
Lab Tech					
Lab report #					

Consec. #					
Date					
Test #					
Chain					
Sta.					
CL Offset					
Lift #					
Material Type					
Proctor #					
Da					
Optimum Moisture					
Plasticy Index					
Liquid Limit					
Density Req.					
Depth					
Moisture Standard.					
Density Standard					
Density Count					
Moisture Count					
Wet Density					
Dry Density					
% Moisture					
% Compaction					
Pass / fail					
Inspector					
Lab Tech					
Lab report #					

Consec. #						
Date						
Test #						
Chain						
Sta.						
CL Offset						
Lift #						
Material Type						
Proctor #						
Da						
Optimum Moisture						
Plasticy Index						
Liquid Limit						
Density Req.						
Depth						
Moisture Standard.						
Density Standard						
Density Count						
Moisture Count						
Wet Density						
Dry Density						
% Moisture						
% Compaction						
Pass / fail						
Inspector						
Lab Tech						
Lab report #						

Consec. #						
Date						
Test #						
Chain						
Sta.						
CL Offset						
Lift #						
Material Type						
Proctor #						
Da						
Optimum Moisture						
Plasticy Index						
Liquid Limit						
Density Req.						
Depth						
Moisture Standard.						
Density Standard						
Density Count						
Moisture Count						
Wet Density						
Dry Density						
% Moisture						
% Compaction						
Pass / fail						
Inspector						
Lab Tech						
Lab report #						

Consec. #					
Date					
Test #					
Chain					
Sta.					
CL Offset					
Lift #					
Material Type					
Proctor #					
Da					
Optimum Moisture					
Plasticy Index					
Liquid Limit					
Density Req.					
Depth					
Moisture Standard.					
Density Standard					
Density Count					
Moisture Count					
Wet Density					
Dry Density					
% Moisture					
% Compaction					
Pass / fail					
Inspector					
Lab Tech					
Lab report #					

Consec. #						
Date						
Test #						
Chain						
Sta.						
CL Offset						
Lift #						
Material Type						
Proctor #						
Da						
Optimum Moisture						
Plasticy Index						
Liquid Limit						
Density Req.						
Depth						
Moisture Standard.						
Density Standard						
Density Count						
Moisture Count						
Wet Density						
Dry Density						
% Moisture						
% Compaction						
Pass / fail						
Inspector						
Lab Tech						
Lab report #						

Consec. #					
Date					
Test #					
Chain					
Sta.					
CL Offset					
Lift #					
Material Type					
Proctor #					
Da					
Optimum Moisture					
Plasticy Index					
Liquid Limit					
Density Req.					
Depth					
Moisture Standard.					
Density Standard					
Density Count					
Moisture Count					
Wet Density					
Dry Density					
% Moisture					
% Compaction					
Pass / fail					
Inspector					
Lab Tech					
Lab report #					

Consec. #						
Date						
Test #						
Chain						
Sta.						
CL Offset						
Lift #						
Material Type						
Proctor #						
Da						
Optimum Moisture						
Plasticy Index						
Liquid Limit						
Density Req.						
Depth						
Moisture Standard.						
Density Standard						
Density Count						
Moisture Count						
Wet Density						
Dry Density						
% Moisture						
% Compaction						
Pass / fail						
Inspector						
Lab Tech						
Lab report #						

Consec. #					
Date					
Test #					
Chain					
Sta.					
CL Offset					
Lift #					
Material Type					
Proctor #					
Da					
Optimum Moisture					
Plasticy Index					
Liquid Limit					
Density Req.					
Depth					
Moisture Standard.					
Density Standard					
Density Count					
Moisture Count					
Wet Density					
Dry Density					
% Moisture					
% Compaction					
Pass / fail					
Inspector					
Lab Tech					
Lab report #					

Consec. #					
Date					
Test #					
Chain					
Sta.					
CL Offset					
Lift #					
Material Type					
Proctor #					
Da					
Optimum Moisture					
Plasticy Index					
Liquid Limit					
Density Req.					
Depth					
Moisture Standard.					
Density Standard					
Density Count					
Moisture Count					
Wet Density					
Dry Density					
% Moisture					
% Compaction					
Pass / fail					
Inspector					
Lab Tech					
Lab report #					

Consec. #						
Date						
Test #						
Chain						
Sta.						
CL Offset						
Lift #						
Material Type						
Proctor #						
Da						
Optimum Moisture						
Plasticy Index						
Liquid Limit						
Density Req.						
Depth						
Moisture Standard.						
Density Standard						
Density Count						
Moisture Count						
Wet Density						
Dry Density						
% Moisture						
% Compaction						
Pass / fail						
Inspector						
Lab Tech						
Lab report #						

Consec. #						
Date						
Test #						
Chain						
Sta.						
CL Offset						
Lift #						
Material Type						
Proctor #						
Da						
Optimum Moisture						
Plasticy Index						
Liquid Limit						
Density Req.						
Depth						
Moisture Standard.						
Density Standard						
Density Count						
Moisture Count						
Wet Density						
Dry Density						
% Moisture						
% Compaction						
Pass / fail						
Inspector						
Lab Tech						
Lab report #						

Consec. #						
Date						
Test #						
Chain						
Sta.						
CL Offset						
Lift #						
Material Type						
Proctor #						
Da						
Optimum Moisture						
Plasticy Index						
Liquid Limit						
Density Req.						
Depth						
Moisture Standard.						
Density Standard						
Density Count						
Moisture Count						
Wet Density						
Dry Density						
% Moisture						
% Compaction						
Pass / fail						
Inspector						
Lab Tech						
Lab report #						

Consec. #					
Date					
Test #					
Chain					
Sta.					
CL Offset					
Lift #					
Material Type					
Proctor #					
Da					
Optimum Moisture					
Plasticy Index					
Liquid Limit					
Density Req.					
Depth					
Moisture Standard.					
Density Standard					
Density Count					
Moisture Count					
Wet Density					
Dry Density					
% Moisture					
% Compaction					
Pass / fail					
Inspector					
Lab Tech					
Lab report #					

Consec. #						
Date						
Test #						
Chain						
Sta.						
CL Offset						
Lift #						
Material Type						
Proctor #						
Da						
Optimum Moisture						
Plasticy Index						
Liquid Limit						
Density Req.						
Depth						
Moisture Standard.						
Density Standard						
Density Count						
Moisture Count						
Wet Density						
Dry Density						
% Moisture						
% Compaction						
Pass / fail						
Inspector						
Lab Tech						
Lab report #						

Consec. #						
Date						
Test #						
Chain						
Sta.						
CL Offset						
Lift #						
Material Type						
Proctor #						
De						
Optimum Moisture						
Plasticy Index						
Liquid Limit						
Density Req.						
Depth						
Moisture Standard.						
Density Standard						
Density Count						
Moisture Count						
Wet Density						
Dry Density						
% Moisture						
% Compaction						
Pass / fail						
Inspector						
Lab Tech						
Lab report #						

Consec. #						
Date						
Test #						
Chain						
Sta.						
CL Offset						
Lift #						
Material Type						
Proctor #						
Da						
Optimum Moisture						
Plasticy Index						
Liquid Limit						
Density Req.						
Depth						
Moisture Standard.						
Density Standard						
Density Count						
Moisture Count						
Wet Density						
Dry Density						
% Moisture						
% Compaction						
Pass / fail						
Inspector						
Lab Tech						
Lab report #						

Consec. #					
Date					
Test #					
Chain					
Sta.					
CL Offset					
Lift #					
Material Type					
Proctor #					
Da					
Optimum Moisture					
Plasticy Index					
Liquid Limit					
Density Req.					
Depth					
Moisture Standard.					
Density Standard					
Density Count					
Moisture Count					
Wet Density					
Dry Density					
% Moisture					
% Compaction					
Pass / fail					
Inspector					
Lab Tech					
Lab report #					

Consec. #					
Date					
Test #					
Chain					
Sta.					
CL Offset					
Lift #					
Material Type					
Proctor #					
Da					
Optimum Moisture					
Plasticy Index					
Liquid Limit					
Density Req.					
Depth					
Moisture Standard.					
Density Standard					
Density Count					
Moisture Count					
Wet Density					
Dry Density					
% Moisture					
% Compaction					
Pass / fail					
Inspector					
Lab Tech					
Lab report #					

Consec. #						
Date						
Test #						
Chain						
Sta.						
CL Offset						
Lift #						
Material Type						
Proctor #						
Da						
Optimum Moisture						
Plasticy Index						
Liquid Limit						
Density Req.						
Depth						
Moisture Standard.						
Density Standard						
Density Count						
Moisture Count						
Wet Density						
Dry Density						
% Moisture						
% Compaction						
Pass / fail						
Inspector						
Lab Tech						
Lab report #						

Consec. #						
Date						
Test #						
Chain						
Sta.						
CL Offset						
Lift #						
Material Type						
Proctor #						
Da						
Optimum Moisture						
Plasticy Index						
Liquid Limit						
Density Req.						
Depth						
Moisture Standard.						
Density Standard						
Density Count						
Moisture Count						
Wet Density						
Dry Density						
% Moisture						
% Compaction						
Pass / fail						
Inspector						
Lab Tech						
Lab report #						

Consec. #					
Date					
Test #					
Chain					
Sta.					
CL Offset					
Lift #					
Material Type					
Proctor #					
Da					
Optimum Moisture					
Plasticy Index					
Liquid Limit					
Density Req.					
Depth					
Moisture Standard.					
Density Standard					
Density Count					
Moisture Count					
Wet Density					
Dry Density					
% Moisture					
% Compaction					
Pass / fail					
Inspector					
Lab Tech					
Lab report #					

Consec. #						
Date						
Test #						
Chain						
Sta.						
CL Offset						
Lift #						
Material Type						
Proctor #						
Da						
Optimum Moisture						
Plasticy Index						
Liquid Limit						
Density Req.						
Depth						
Moisture Standard.						
Density Standard						
Density Count						
Moisture Count						
Wet Density						
Dry Density						
% Moisture						
% Compaction						
Pass / fail						
Inspector						
Lab Tech						
Lab report #						

Consec. #						
Date						
Test #						
Chain						
Sta.						
CL Offset						
Lift #						
Material Type						
Proctor #						
Da						
Optimum Moisture						
Plasticy Index						
Liquid Limit						
Density Req.						
Depth						
Moisture Standard.						
Density Standard						
Density Count						
Moisture Count						
Wet Density						
Dry Density						
% Moisture						
% Compaction						
Pass / fail						
Inspector						
Lab Tech						
Lab report #						

Consec. #					
Date					
Test #					
Chain					
Sta.					
CL Offset					
Lift #					
Material Type					
Proctor #					
Da					
Optimum Moisture					
Plasticy Index					
Liquid Limit					
Density Req.					
Depth					
Moisture Standard.					
Density Standard					
Density Count					
Moisture Count					
Wet Density					
Dry Density					
% Moisture					
% Compaction					
Pass / fail					
Inspector					
Lab Tech					
Lab report #					

Consec. #						
Date						
Test #						
Chain						
Sta.						
CL Offset						
Lift #						
Material Type						
Proctor #						
Da						
Optimum Moisture						
Plasticy Index						
Liquid Limit						
Density Req.						
Depth						
Moisture Standard.						
Density Standard						
Density Count						
Moisture Count						
Wet Density						
Dry Density						
% Moisture						
% Compaction						
Pass / fail						
Inspector						
Lab Tech						
Lab report #						

Consec. #					
Date					
Test #					
Chain					
Sta.					
CL Offset					
Lift #					
Material Type					
Proctor #					
Da					
Optimum Moisture					
Plasticy Index					
Liquid Limit					
Density Req.					
Depth					
Moisture Standard.					
Density Standard					
Density Count					
Moisture Count					
Wet Density					
Dry Density					
% Moisture					
% Compaction					
Pass / fail					
Inspector					
Lab Tech					
Lab report #					

Consec. #						
Date						
Test #						
Chain						
Sta.						
CL Offset						
Lift #						
Material Type						
Proctor #						
Da						
Optimum Moisture						
Plasticy Index						
Liquid Limit						
Density Req.						
Depth						
Moisture Standard.						
Density Standard						
Density Count						
Moisture Count						
Wet Density						
Dry Density						
% Moisture						
% Compaction						
Pass / fail						
Inspector						
Lab Tech						
Lab report #						

Consec. #						
Date						
Test #						
Chain						
Sta.						
CL Offset						
Lift #						
Material Type						
Proctor #						
Da						
Optimum Moisture						
Plasticy Index						
Liquid Limit						
Density Req.						
Depth						
Moisture Standard.						
Density Standard						
Density Count						
Moisture Count						
Wet Density						
Dry Density						
% Moisture						
% Compaction						
Pass / fail						
Inspector						
Lab Tech						
Lab report #						

Consec. #					
Date					
Test #					
Chain					
Sta.					
CL Offset					
Lift #					
Material Type					
Proctor #					
Da					
Optimum Moisture					
Plasticy Index					
Liquid Limit					
Density Req.					
Depth					
Moisture Standard.					
Density Standard					
Density Count					
Moisture Count					
Wet Density					
Dry Density					
% Moisture					
% Compaction					
Pass / fail					
Inspector					
Lab Tech					
Lab report #					

Consec. #					
Date					
Test #					
Chain					
Sta.					
CL Offset					
Lift #					
Material Type					
Proctor #					
Da					
Optimum Moisture					
Plasticy Index					
Liquid Limit					
Density Req.					
Depth					
Moisture Standard.					
Density Standard					
Density Count					
Moisture Count					
Wet Density					
Dry Density					
% Moisture					
% Compaction					
Pass / fail					
Inspector					
Lab Tech					
Lab report #					

Consec. #						
Date						
Test #						
Chain						
Sta.						
CL Offset						
Lift #						
Material Type						
Proctor #						
Da						
Optimum Moisture						
Plasticy Index						
Liquid Limit						
Density Req.						
Depth						
Moisture Standard.						
Density Standard						
Density Count						
Moisture Count						
Wet Density						
Dry Density						
% Moisture						
% Compaction						
Pass / fail						
Inspector						
Lab Tech						
Lab report #						

Consec. #						
Date						
Test #						
Chain						
Sta.						
CL Offset						
Lift #						
Material Type						
Proctor #						
Da						
Optimum Moisture						
Plasticy Index						
Liquid Limit						
Density Req.						
Depth						
Moisture Standard.						
Density Standard						
Density Count						
Moisture Count						
Wet Density						
Dry Density						
% Moisture						
% Compaction						
Pass / fail						
Inspector						
Lab Tech						
Lab report #						

Consec. #						
Date						
Test #						
Chain						
Sta.						
CL Offset						
Lift #						
Material Type						
Proctor #						
Da						
Optimum Moisture						
Plasticy Index						
Liquid Limit						
Density Req.						
Depth						
Moisture Standard.						
Density Standard						
Density Count						
Moisture Count						
Wet Density						
Dry Density						
% Moisture						
% Compaction						
Pass / fail						
Inspector						
Lab Tech						
Lab report #						

Consec. #						
Date						
Test #						
Chain						
Sta.						
CL Offset						
Lift #						
Material Type						
Proctor #						
Da						
Optimum Moisture						
Plasticy Index						
Liquid Limit						
Density Req.						
Depth						
Moisture Standard.						
Density Standard						
Density Count						
Moisture Count						
Wet Density						
Dry Density						
% Moisture						
% Compaction						
Pass / fail						
Inspector						
Lab Tech						
Lab report #						

Consec. #						
Date						
Test #						
Chain						
Sta.						
CL Offset						
Lift #						
Material Type						
Proctor #						
Da						
Optimum Moisture						
Plasticy Index						
Liquid Limit						
Density Req.						
Depth						
Moisture Standard.						
Density Standard						
Density Count						
Moisture Count						
Wet Density						
Dry Density						
% Moisture						
% Compaction						
Pass / fail						
Inspector						
Lab Tech						
Lab report #						

Consec. #					
Date					
Test #					
Chain					
Sta.					
CL Offset					
Lift #					
Material Type					
Proctor #					
Da					
Optimum Moisture					
Plasticy Index					
Liquid Limit					
Density Req.					
Depth					
Moisture Standard.					
Density Standard					
Density Count					
Moisture Count					
Wet Density					
Dry Density					
% Moisture					
% Compaction					
Pass / fail					
Inspector					
Lab Tech					
Lab report #					

Consec. #						
Date						
Test #						
Chain						
Sta.						
CL Offset						
Lift #						
Material Type						
Proctor #						
Da						
Optimum Moisture						
Plasticy Index						
Liquid Limit						
Density Req.						
Depth						
Moisture Standard.						
Density Standard						
Density Count						
Moisture Count						
Wet Density						
Dry Density						
% Moisture						
% Compaction						
Pass / fail						
Inspector						
Lab Tech						
Lab report #						

Consec. #						
Date						
Test #						
Chain						
Sta.						
CL Offset						
Lift #						
Material Type						
Proctor #						
Da						
Optimum Moisture						
Plasticy Index						
Liquid Limit						
Density Req.						
Depth						
Moisture Standard.						
Density Standard						
Density Count						
Moisture Count						
Wet Density						
Dry Density						
% Moisture						
% Compaction						
Pass / fail						
Inspector						
Lab Tech						
Lab report #						

Consec. #						
Date						
Test #						
Chain						
Sta.						
CL Offset						
Lift #						
Material Type						
Proctor #						
Da						
Optimum Moisture						
Plasticy Index						
Liquid Limit						
Density Req.						
Depth						
Moisture Standard.						
Density Standard						
Density Count						
Moisture Count						
Wet Density						
Dry Density						
% Moisture						
% Compaction						
Pass / fail						
Inspector						
Lab Tech						
Lab report #						

Consec. #					
Date					
Test #					
Chain					
Sta.					
CL Offset					
Lift #					
Material Type					
Proctor #					
Da					
Optimum Moisture					
Plasticy Index					
Liquid Limit					
Density Req.					
Depth					
Moisture Standard.					
Density Standard					
Density Count					
Moisture Count					
Wet Density					
Dry Density					
% Moisture					
% Compaction					
Pass /fail					
Inspector					
Lab Tech					
Lab report #					

Consec. #						
Date						
Test #						
Chain						
Sta.						
CL Offset						
Lift #						
Material Type						
Proctor #						
Da						
Optimum Moisture						
Plasticy Index						
Liquid Limit						
Density Req.						
Depth						
Moisture Standard.						
Density Standard						
Density Count						
Moisture Count						
Wet Density						
Dry Density						
% Moisture						
% Compaction						
Pass / fail						
Inspector						
Lab Tech						
Lab report #						

Consec. #					
Date					
Test #					
Chain					
Sta.					
CL Offset					
Lift #					
Material Type					
Proctor #					
Da					
Optimum Moisture					
Plasticy Index					
Liquid Limit					
Density Req.					
Depth					
Moisture Standard.					
Density Standard					
Density Count					
Moisture Count					
Wet Density					
Dry Density					
% Moisture					
% Compaction					
Pass / fail					
Inspector					
Lab Tech					
Lab report #					

Consec. #						
Date						
Test #						
Chain						
Sta.						
CL Offset						
Lift #						
Material Type						
Proctor #						
Da						
Optimum Moisture						
Plasticy Index						
Liquid Limit						
Density Req.						
Depth						
Moisture Standard.						
Density Standard						
Density Count						
Moisture Count						
Wet Density						
Dry Density						
% Moisture						
% Compaction						
Pass / fail						
Inspector						
Lab Tech						
Lab report #						

Consec. #					
Date					
Test #					
Chain					
Sta.					
CL Offset					
Lift #					
Material Type					
Proctor #					
Da					
Optimum Moisture					
Plasticy Index					
Liquid Limit					
Density Req.					
Depth					
Moisture Standard.					
Density Standard					
Density Count					
Moisture Count					
Wet Density					
Dry Density					
% Moisture					
% Compaction					
Pass / fail					
Inspector					
Lab Tech					
Lab report #					

Consec. #						
Date						
Test #						
Chain						
Sta.						
CL Offset						
Lift #						
Material Type						
Proctor #						
Da						
Optimum Moisture						
Plasticy Index						
Liquid Limit						
Density Req.						
Depth						
Moisture Standard.						
Density Standard						
Density Count						
Moisture Count						
Wet Density						
Dry Density						
% Moisture						
% Compaction						
Pass / fail						
Inspector						
Lab Tech						
Lab report #						

Consec. #					
Date					
Test #					
Chain					
Sta.					
CL Offset					
Lift #					
Material Type					
Proctor #					
Da					
Optimum Moisture					
Plasticy Index					
Liquid Limit					
Density Req.					
Depth					
Moisture Standard.					
Density Standard					
Density Count					
Moisture Count					
Wet Density					
Dry Density					
% Moisture					
% Compaction					
Pass / fail					
Inspector					
Lab Tech					
Lab report #					

Consec. #					
Date					
Test #					
Chain					
Sta.					
CL Offset					
Lift #					
Material Type					
Proctor #					
Da					
Optimum Moisture					
Plasticy Index					
Liquid Limit					
Density Req.					
Depth					
Moisture Standard.					
Density Standard					
Density Count					
Moisture Count					
Wet Density					
Dry Density					
% Moisture					
% Compaction					
Pass / fail					
Inspector					
Lab Tech					
Lab report #					

Consec. #					
Date					
Test #					
Chain					
Sta.					
CL Offset					
Lift #					
Material Type					
Proctor #					
Da					
Optimum Moisture					
Plasticy Index					
Liquid Limit					
Density Req.					
Depth					
Moisture Standard.					
Density Standard					
Density Count					
Moisture Count					
Wet Density					
Dry Density					
% Moisture					
% Compaction					
Pass / fail					
Inspector					
Lab Tech					
Lab report #					

Consec. #						
Date						
Test #						
Chain						
Sta.						
CL Offset						
Lift #						
Material Type						
Proctor #						
Da						
Optimum Moisture						
Plasticy Index						
Liquid Limit						
Density Req.						
Depth						
Moisture Standard.						
Density Standard						
Density Count						
Moisture Count						
Wet Density						
Dry Density						
% Moisture						
% Compaction						
Pass / fail						
Inspector						
Lab Tech						
Lab report #						

Consec. #						
Date						
Test #						
Chain						
Sta.						
CL Offset						
Lift #						
Material Type						
Proctor #						
Da						
Optimum Moisture						
Plasticy Index						
Liquid Limit						
Density Req.						
Depth						
Moisture Standard.						
Density Standard						
Density Count						
Moisture Count						
Wet Density						
Dry Density						
% Moisture						
% Compaction						
Pass / fail						
Inspector						
Lab Tech						
Lab report #						

Consec. #						
Date						
Test #						
Chain						
Sta.						
CL Offset						
Lift #						
Material Type						
Proctor #						
Da						
Optimum Moisture						
Plasticy Index						
Liquid Limit						
Density Req.						
Depth						
Moisture Standard.						
Density Standard						
Density Count						
Moisture Count						
Wet Density						
Dry Density						
% Moisture						
% Compaction						
Pass / fail						
Inspector						
Lab Tech						
Lab report #						

Consec. #					
Date					
Test #					
Chain					
Sta.					
CL Offset					
Lift #					
Material Type					
Proctor #					
Da					
Optimum Moisture					
Plasticy Index					
Liquid Limit					
Density Req.					
Depth					
Moisture Standard.					
Density Standard					
Density Count					
Moisture Count					
Wet Density					
Dry Density					
% Moisture					
% Compaction					
Pass / fail					
Inspector					
Lab Tech					
Lab report #					

Consec. #						
Date						
Test #						
Chain						
Sta.						
CL Offset						
Lift #						
Material Type						
Proctor #						
Da						
Optimum Moisture						
Plasticy Index						
Liquid Limit						
Density Req.						
Depth						
Moisture Standard.						
Density Standard						
Density Count						
Moisture Count						
Wet Density						
Dry Density						
% Moisture						
% Compaction						
Pass / fail						
Inspector						
Lab Tech						
Lab report #						

Consec. #					
Date					
Test #					
Chain					
Sta.					
CL Offset					
Lift #					
Material Type					
Proctor #					
Da					
Optimum Moisture					
Plasticy Index					
Liquid Limit					
Density Req.					
Depth					
Moisture Standard.					
Density Standard					
Density Count					
Moisture Count					
Wet Density					
Dry Density					
% Moisture					
% Compaction					
Pass / fail					
Inspector					
Lab Tech					
Lab report #					

Consec. #					
Date					
Test #					
Chain					
Sta.					
CL Offset					
Lift #					
Material Type					
Proctor #					
Da					
Optimum Moisture					
Plasticy Index					
Liquid Limit					
Density Req.					
Depth					
Moisture Standard.					
Density Standard					
Density Count					
Moisture Count					
Wet Density					
Dry Density					
% Moisture					
% Compaction					
Pass / fail					
Inspector					
Lab Tech					
Lab report #					

Consec. #						
Date						
Test #						
Chain						
Sta.						
CL Offset						
Lift #						
Material Type						
Proctor #						
Da						
Optimum Moisture						
Plasticy Index						
Liquid Limit						
Density Req.						
Depth						
Moisture Standard.						
Density Standard						
Density Count						
Moisture Count						
Wet Density						
Dry Density						
% Moisture						
% Compaction						
Pass / fail						
Inspector						
Lab Tech						
Lab report #						

Consec. #						
Date						
Test #						
Chain						
Sta.						
CL Offset						
Lift #						
Material Type						
Proctor #						
Da						
Optimum Moisture						
Plasticy Index						
Liquid Limit						
Density Req.						
Depth						
Moisture Standard.						
Density Standard						
Density Count						
Moisture Count						
Wet Density						
Dry Density						
% Moisture						
% Compaction						
Pass / fail						
Inspector						
Lab Tech						
Lab report #						

Consec. #					
Date					
Test #					
Chain					
Sta.					
CL Offset					
Lift #					
Material Type					
Proctor #					
Da					
Optimum Moisture					
Plasticy Index					
Liquid Limit					
Density Req.					
Depth					
Moisture Standard.					
Density Standard					
Density Count					
Moisture Count					
Wet Density					
Dry Density					
% Moisture					
% Compaction					
Pass / fail					
Inspector					
Lab Tech					
Lab report #					

Consec. #					
Date					
Test #					
Chain					
Sta.					
CL Offset					
Lift #					
Material Type					
Proctor #					
Da					
Optimum Moisture					
Plasticy Index					
Liquid Limit					
Density Req.					
Depth					
Moisture Standard.					
Density Standard					
Density Count					
Moisture Count					
Wet Density					
Dry Density					
% Moisture					
% Compaction					
Pass / fail					
Inspector					
Lab Tech					
Lab report #					

Consec. #					
Date					
Test #					
Chain					
Sta.					
CL Offset					
Lift #					
Material Type					
Proctor #					
Da					
Optimum Moisture					
Plasticy Index					
Liquid Limit					
Density Req.					
Depth					
Moisture Standard.					
Density Standard					
Density Count					
Moisture Count					
Wet Density					
Dry Density					
% Moisture					
% Compaction					
Pass / fail					
Inspector					
Lab Tech					
Lab report #					

Consec. #						
Date						
Test #						
Chain						
Sta.						
CL Offset						
Lift #						
Material Type						
Proctor #						
Da						
Optimum Moisture						
Plasticy Index						
Liquid Limit						
Density Req.						
Depth						
Moisture Standard.						
Density Standard						
Density Count						
Moisture Count						
Wet Density						
Dry Density						
% Moisture						
% Compaction						
Pass / fail						
Inspector						
Lab Tech						
Lab report #						

Consec. #						
Date						
Test #						
Chain						
Sta.						
CL Offset						
Lift #						
Material Type						
Proctor #						
Da						
Optimum Moisture						
Plasticy Index						
Liquid Limit						
Density Req.						
Depth						
Moisture Standard.						
Density Standard						
Density Count						
Moisture Count						
Wet Density						
Dry Density						
% Moisture						
% Compaction						
Pass / fail						
Inspector						
Lab Tech						
Lab report #						

Consec. #						
Date						
Test #						
Chain						
Sta.						
CL Offset						
Lift #						
Material Type						
Proctor #						
Da						
Optimum Moisture						
Plasticy Index						
Liquid Limit						
Density Req.						
Depth						
Moisture Standard.						
Density Standard						
Density Count						
Moisture Count						
Wet Density						
Dry Density						
% Moisture						
% Compaction						
Pass / fail						
Inspector						
Lab Tech						
Lab report #						

Consec. #					
Date					
Test #					
Chain					
Sta.					
CL Offset					
Lift #					
Material Type					
Proctor #					
Da					
Optimum Moisture					
Plasticy Index					
Liquid Limit					
Density Req.					
Depth					
Moisture Standard.					
Density Standard					
Density Count					
Moisture Count					
Wet Density					
Dry Density					
% Moisture					
% Compaction					
Pass / fail					
Inspector					
Lab Tech					
Lab report #					

Consec. #						
Date						
Test #						
Chain						
Sta.						
CL Offset						
Lift #						
Material Type						
Proctor #						
Da						
Optimum Moisture						
Plasticy Index						
Liquid Limit						
Density Req.						
Depth						
Moisture Standard.						
Density Standard						
Density Count						
Moisture Count						
Wet Density						
Dry Density						
% Moisture						
% Compaction						
Pass / fail						
Inspector						
Lab Tech						
Lab report #						

Consec. #						
Date						
Test #						
Chain						
Sta.						
CL Offset						
Lift #						
Material Type						
Proctor #						
Da						
Optimum Moisture						
Plasticy Index						
Liquid Limit						
Density Req.						
Depth						
Moisture Standard.						
Density Standard						
Density Count						
Moisture Count						
Wet Density						
Dry Density						
% Moisture						
% Compaction						
Pass / fail						
Inspector						
Lab Tech						
Lab report #						

Consec. #						
Date						
Test #						
Chain						
Sta.						
CL Offset						
Lift #						
Material Type						
Proctor #						
Da						
Optimum Moisture						
Plasticy Index						
Liquid Limit						
Density Req.						
Depth						
Moisture Standard.						
Density Standard						
Density Count						
Moisture Count						
Wet Density						
Dry Density						
% Moisture						
% Compaction						
Pass / fail						
Inspector						
Lab Tech						
Lab report #						

Consec. #						
Date						
Test #						
Chain						
Sta.						
CL Offset						
Lift #						
Material Type						
Proctor #						
Da						
Optimum Moisture						
Plasticy Index						
Liquid Limit						
Density Req.						
Depth						
Moisture Standard.						
Density Standard						
Density Count						
Moisture Count						
Wet Density						
Dry Density						
% Moisture						
% Compaction						
Pass / fail						
Inspector						
Lab Tech						
Lab report #						

Consec. #						
Date						
Test #						
Chain						
Sta.						
CL Offset						
Lift #						
Material Type						
Proctor #						
Da						
Optimum Moisture						
Plasticy Index						
Liquid Limit						
Density Req.						
Depth						
Moisture Standard.						
Density Standard						
Density Count						
Moisture Count						
Wet Density						
Dry Density						
% Moisture						
% Compaction						
Pass / fail						
Inspector						
Lab Tech						
Lab report #						

Consec. #					
Date					
Test #					
Chain					
Sta.					
CL Offset					
Lift #					
Material Type					
Proctor #					
Da					
Optimum Moisture					
Plasticy Index					
Liquid Limit					
Density Req.					
Depth					
Moisture Standard.					
Density Standard					
Density Count					
Moisture Count					
Wet Density					
Dry Density					
% Moisture					
% Compaction					
Pass / fail					
Inspector					
Lab Tech					
Lab report #					

Consec. #					
Date					
Test #					
Chain					
Sta.					
CL Offset					
Lift #					
Material Type					
Proctor #					
Da					
Optimum Moisture					
Plasticy Index					
Liquid Limit					
Density Req.					
Depth					
Moisture Standard.					
Density Standard					
Density Count					
Moisture Count					
Wet Density					
Dry Density					
% Moisture					
% Compaction					
Pass / fail					
Inspector					
Lab Tech					
Lab report #					

Consec. #					
Date					
Test #					
Chain					
Sta.					
CL Offset					
Lift #					
Material Type					
Proctor #					
Da					
Optimum Moisture					
Plasticy Index					
Liquid Limit					
Density Req.					
Depth					
Moisture Standard.					
Density Standard					
Density Count					
Moisture Count					
Wet Density					
Dry Density					
% Moisture					
% Compaction					
Pass / fail					
Inspector					
Lab Tech					
Lab report #					

Consec. #						
Date						
Test #						
Chain						
Sta.						
CL Offset						
Lift #						
Material Type						
Proctor #						
Da						
Optimum Moisture						
Plasticy Index						
Liquid Limit						
Density Req.						
Depth						
Moisture Standard.						
Density Standard						
Density Count						
Moisture Count						
Wet Density						
Dry Density						
% Moisture						
% Compaction						
Pass / fail						
Inspector						
Lab Tech						
Lab report #						

Consec. #						
Date						
Test #						
Chain						
Sta.						
CL Offset						
Lift #						
Material Type						
Proctor #						
Da						
Optimum Moisture						
Plasticy Index						
Liquid Limit						
Density Req.						
Depth						
Moisture Standard.						
Density Standard						
Density Count						
Moisture Count						
Wet Density						
Dry Density						
% Moisture						
% Compaction						
Pass / fail						
Inspector						
Lab Tech						
Lab report #						

Consec. #						
Date						
Test #						
Chain						
Sta.						
CL Offset						
Lift #						
Material Type						
Proctor #						
Da						
Optimum Moisture						
Plasticy Index						
Liquid Limit						
Density Req.						
Depth						
Moisture Standard.						
Density Standard						
Density Count						
Moisture Count						
Wet Density						
Dry Density						
% Moisture						
% Compaction						
Pass / fail						
Inspector						
Lab Tech						
Lab report #						

Consec. #						
Date						
Test #						
Chain						
Sta.						
CL Offset						
Lift #						
Material Type						
Proctor #						
Da						
Optimum Moisture						
Plasticy Index						
Liquid Limit						
Density Req.						
Depth						
Moisture Standard.						
Density Standard						
Density Count						
Moisture Count						
Wet Density						
Dry Density						
% Moisture						
% Compaction						
Pass / fail						
Inspector						
Lab Tech						
Lab report #						

Consec. #						
Date						
Test #						
Chain						
Sta.						
CL Offset						
Lift #						
Material Type						
Proctor #						
Da						
Optimum Moisture						
Plasticy Index						
Liquid Limit						
Density Req.						
Depth						
Moisture Standard.						
Density Standard						
Density Count						
Moisture Count						
Wet Density						
Dry Density						
% Moisture						
% Compaction						
Pass / fail						
Inspector						
Lab Tech						
Lab report #						

Consec. #						
Date						
Test #						
Chain						
Sta.						
CL Offset						
Lift #						
Material Type						
Proctor #						
Da						
Optimum Moisture						
Plasticy Index						
Liquid Limit						
Density Req.						
Depth						
Moisture Standard.						
Density Standard						
Density Count						
Moisture Count						
Wet Density						
Dry Density						
% Moisture						
% Compaction						
Pass / fail						
Inspector						
Lab Tech						
Lab report #						

Consec. #						
Date						
Test #						
Chain						
Sta.						
CL Offset						
Lift #						
Material Type						
Proctor #						
Da						
Optimum Moisture						
Plasticy Index						
Liquid Limit						
Density Req.						
Depth						
Moisture Standard.						
Density Standard						
Density Count						
Moisture Count						
Wet Density						
Dry Density						
% Moisture						
% Compaction						
Pass / fail						
Inspector						
Lab Tech						
Lab report #						

Consec. #						
Date						
Test #						
Chain						
Sta.						
CL Offset						
Lift #						
Material Type						
Proctor #						
Da						
Optimum Moisture						
Plasticy Index						
Liquid Limit						
Density Req.						
Depth						
Moisture Standard.						
Density Standard						
Density Count						
Moisture Count						
Wet Density						
Dry Density						
% Moisture						
% Compaction						
Pass / fail						
Inspector						
Lab Tech						
Lab report #						

Consec. #						
Date						
Test #						
Chain						
Sta.						
CL Offset						
Lift #						
Material Type						
Proctor #						
Da						
Optimum Moisture						
Plasticy Index						
Liquid Limit						
Density Req.						
Depth						
Moisture Standard.						
Density Standard						
Density Count						
Moisture Count						
Wet Density						
Dry Density						
% Moisture						
% Compaction						
Pass / fail						
Inspector						
Lab Tech						
Lab report #						

Consec. #						
Date						
Test #						
Chain						
Sta.						
CL Offset						
Lift #						
Material Type						
Proctor #						
Da						
Optimum Moisture						
Plasticy Index						
Liquid Limit						
Density Req.						
Depth						
Moisture Standard.						
Density Standard						
Density Count						
Moisture Count						
Wet Density						
Dry Density						
% Moisture						
% Compaction						
Pass / fail						
Inspector						
Lab Tech						
Lab report #						

Consec. #					
Date					
Test #					
Chain					
Sta.					
CL Offset					
Lift #					
Material Type					
Proctor #					
Da					
Optimum Moisture					
Plasticy Index					
Liquid Limit					
Density Req.					
Depth					
Moisture Standard.					
Density Standard					
Density Count					
Moisture Count					
Wet Density					
Dry Density					
% Moisture					
% Compaction					
Pass / fail					
Inspector					
Lab Tech					
Lab report #					

Consec. #						
Date						
Test #						
Chain						
Sta.						
CL Offset						
Lift #						
Material Type						
Proctor #						
Da						
Optimum Moisture						
Plasticy Index						
Liquid Limit						
Density Req.						
Depth						
Moisture Standard.						
Density Standard						
Density Count						
Moisture Count						
Wet Density						
Dry Density						
% Moisture						
% Compaction						
Pass / fail						
Inspector						
Lab Tech						
Lab report #						

Consec. #						
Date						
Test #						
Chain						
Sta.						
CL Offset						
Lift #						
Material Type						
Proctor #						
Da						
Optimum Moisture						
Plasticy Index						
Liquid Limit						
Density Req.						
Depth						
Moisture Standard.						
Density Standard						
Density Count						
Moisture Count						
Wet Density						
Dry Density						
% Moisture						
% Compaction						
Pass / fail						
Inspector						
Lab Tech						
Lab report #						

Consec. #					
Date					
Test #					
Chain					
Sta.					
CL Offset					
Lift #					
Material Type					
Proctor #					
Da					
Optimum Moisture					
Plasticy Index					
Liquid Limit					
Density Req.					
Depth					
Moisture Standard.					
Density Standard					
Density Count					
Moisture Count					
Wet Density					
Dry Density					
% Moisture					
% Compaction					
Pass / fail					
Inspector					
Lab Tech					
Lab report #					

Consec. #						
Date						
Test #						
Chain						
Sta.						
CL Offset						
Lift #						
Material Type						
Proctor #						
Da						
Optimum Moisture						
Plasticy Index						
Liquid Limit						
Density Req.						
Depth						
Moisture Standard.						
Density Standard						
Density Count						
Moisture Count						
Wet Density						
Dry Density						
% Moisture						
% Compaction						
Pass / fail						
Inspector						
Lab Tech						
Lab report #						

Consec. #						
Date						
Test #						
Chain						
Sta.						
CL Offset						
Lift #						
Material Type						
Proctor #						
Da						
Optimum Moisture						
Plasticy Index						
Liquid Limit						
Density Req.						
Depth						
Moisture Standard.						
Density Standard						
Density Count						
Moisture Count						
Wet Density						
Dry Density						
% Moisture						
% Compaction						
Pass / fail						
Inspector						
Lab Tech						
Lab report #						

Consec. #						
Date						
Test #						
Chain						
Sta.						
CL Offset						
Lift #						
Material Type						
Proctor #						
Da						
Optimum Moisture						
Plasticy Index						
Liquid Limit						
Density Req.						
Depth						
Moisture Standard.						
Density Standard						
Density Count						
Moisture Count						
Wet Density						
Dry Density						
% Moisture						
% Compaction						
Pass / fail						
Inspector						
Lab Tech						
Lab report #						

Consec. #						
Date						
Test #						
Chain						
Sta.						
CL Offset						
Lift #						
Material Type						
Proctor #						
Da						
Optimum Moisture						
Plasticy Index						
Liquid Limit						
Density Req.						
Depth						
Moisture Standard.						
Density Standard						
Density Count						
Moisture Count						
Wet Density						
Dry Density						
% Moisture						
% Compaction						
Pass / fail						
Inspector						
Lab Tech						
Lab report #						

Consec. #						
Date						
Test #						
Chain						
Sta.						
CL Offset						
Lift #						
Material Type						
Proctor #						
Da						
Optimum Moisture						
Plasticy Index						
Liquid Limit						
Density Req.						
Depth						
Moisture Standard.						
Density Standard						
Density Count						
Moisture Count						
Wet Density						
Dry Density						
% Moisture						
% Compaction						
Pass / fail						
Inspector						
Lab Tech						
Lab report #						

Consec. #					
Date					
Test #					
Chain					
Sta.					
CL Offset					
Lift #					
Material Type					
Proctor #					
Da					
Optimum Moisture					
Plasticy Index					
Liquid Limit					
Density Req.					
Depth					
Moisture Standard.					
Density Standard					
Density Count					
Moisture Count					
Wet Density					
Dry Density					
% Moisture					
% Compaction					
Pass / fail					
Inspector					
Lab Tech					
Lab report #					

Consec. #						
Date						
Test #						
Chain						
Sta.						
CL Offset						
Lift #						
Material Type						
Proctor #						
Da						
Optimum Moisture						
Plasticy Index						
Liquid Limit						
Density Req.						
Depth						
Moisture Standard.						
Density Standard						
Density Count						
Moisture Count						
Wet Density						
Dry Density						
% Moisture						
% Compaction						
Pass / fail						
Inspector						
Lab Tech						
Lab report #						

Consec. #					
Date					
Test #					
Chain					
Sta.					
CL Offset					
Lift #					
Material Type					
Proctor #					
Da					
Optimum Moisture					
Plasticy Index					
Liquid Limit					
Density Req.					
Depth					
Moisture Standard.					
Density Standard					
Density Count					
Moisture Count					
Wet Density					
Dry Density					
% Moisture					
% Compaction					
Pass / fail					
Inspector					
Lab Tech					
Lab report #					

Consec. #						
Date						
Test #						
Chain						
Sta.						
CL Offset						
Lift #						
Material Type						
Proctor #						
Da						
Optimum Moisture						
Plasticy Index						
Liquid Limit						
Density Req.						
Depth						
Moisture Standard.						
Density Standard						
Density Count						
Moisture Count						
Wet Density						
Dry Density						
% Moisture						
% Compaction						
Pass / fail						
Inspector						
Lab Tech						
Lab report #						

Consec. #					
Date					
Test #					
Chain					
Sta.					
CL Offset					
Lift #					
Material Type					
Proctor #					
Da					
Optimum Moisture					
Plasticy Index					
Liquid Limit					
Density Req.					
Depth					
Moisture Standard.					
Density Standard					
Density Count					
Moisture Count					
Wet Density					
Dry Density					
% Moisture					
% Compaction					
Pass / fail					
Inspector					
Lab Tech					
Lab report #					

Consec. #						
Date						
Test #						
Chain						
Sta.						
CL Offset						
Lift #						
Material Type						
Proctor #						
Da						
Optimum Moisture						
Plasticy Index						
Liquid Limit						
Density Req.						
Depth						
Moisture Standard.						
Density Standard						
Density Count						
Moisture Count						
Wet Density						
Dry Density						
% Moisture						
% Compaction						
Pass / fail						
Inspector						
Lab Tech						
Lab report #						

Consec. #					
Date					
Test #					
Chain					
Sta.					
CL Offset					
Lift #					
Material Type					
Proctor #					
Da					
Optimum Moisture					
Plasticy Index					
Liquid Limit					
Density Req.					
Depth					
Moisture Standard.					
Density Standard					
Density Count					
Moisture Count					
Wet Density					
Dry Density					
% Moisture					
% Compaction					
Pass / fail					
Inspector					
Lab Tech					
Lab report #					

Consec. #						
Date						
Test #						
Chain						
Sta.						
CL Offset						
Lift #						
Material Type						
Proctor #						
Da						
Optimum Moisture						
Plasticy Index						
Liquid Limit						
Density Req.						
Depth						
Moisture Standard.						
Density Standard						
Density Count						
Moisture Count						
Wet Density						
Dry Density						
% Moisture						
% Compaction						
Pass / fail						
Inspector						
Lab Tech						
Lab report #						

Consec. #					
Date					
Test #					
Chain					
Sta.					
CL Offset					
Lift #					
Material Type					
Proctor #					
Da					
Optimum Moisture					
Plasticy Index					
Liquid Limit					
Density Req.					
Depth					
Moisture Standard.					
Density Standard					
Density Count					
Moisture Count					
Wet Density					
Dry Density					
% Moisture					
% Compaction					
Pass /fail					
Inspector					
Lab Tech					
Lab report #					

Consec. #						
Date						
Test #						
Chain						
Sta.						
CL Offset						
Lift #						
Material Type						
Proctor #						
Da						
Optimum Moisture						
Plasticy Index						
Liquid Limit						
Density Req.						
Depth						
Moisture Standard.						
Density Standard						
Density Count						
Moisture Count						
Wet Density						
Dry Density						
% Moisture						
% Compaction						
Pass / fail						
Inspector						
Lab Tech						
Lab report #						

Consec. #					
Date					
Test #					
Chain					
Sta.					
CL Offset					
Lift #					
Material Type					
Proctor #					
Da					
Optimum Moisture					
Plasticy Index					
Liquid Limit					
Density Req.					
Depth					
Moisture Standard.					
Density Standard					
Density Count					
Moisture Count					
Wet Density					
Dry Density					
% Moisture					
% Compaction					
Pass / fail					
Inspector					
Lab Tech					
Lab report #					

Consec. #						
Date						
Test #						
Chain						
Sta.						
CL Offset						
Lift #						
Material Type						
Proctor #						
Da						
Optimum Moisture						
Plasticy Index						
Liquid Limit						
Density Req.						
Depth						
Moisture Standard.						
Density Standard						
Density Count						
Moisture Count						
Wet Density						
Dry Density						
% Moisture						
% Compaction						
Pass / fail						
Inspector						
Lab Tech						
Lab report #						

Consec. #					
Date					
Test #					
Chain					
Sta.					
CL Offset					
Lift #					
Material Type					
Proctor #					
Da					
Optimum Moisture					
Plasticy Index					
Liquid Limit					
Density Req.					
Depth					
Moisture Standard.					
Density Standard					
Density Count					
Moisture Count					
Wet Density					
Dry Density					
% Moisture					
% Compaction					
Pass / fail					
Inspector					
Lab Tech					
Lab report #					

Consec. #					
Date					
Test #					
Chain					
Sta.					
CL Offset					
Lift #					
Material Type					
Proctor #					
Da					
Optimum Moisture					
Plasticy Index					
Liquid Limit					
Density Req.					
Depth					
Moisture Standard.					
Density Standard					
Density Count					
Moisture Count					
Wet Density					
Dry Density					
% Moisture					
% Compaction					
Pass / fail					
Inspector					
Lab Tech					
Lab report #					

Consec. #						
Date						
Test #						
Chain						
Sta.						
CL Offset						
Lift #						
Material Type						
Proctor #						
Da						
Optimum Moisture						
Plasticy Index						
Liquid Limit						
Density Req.						
Depth						
Moisture Standard.						
Density Standard						
Density Count						
Moisture Count						
Wet Density						
Dry Density						
% Moisture						
% Compaction						
Pass / fail						
Inspector						
Lab Tech						
Lab report #						

Consec. #					
Date					
Test #					
Chain					
Sta.					
CL Offset					
Lift #					
Material Type					
Proctor #					
Da					
Optimum Moisture					
Plasticy Index					
Liquid Limit					
Density Req.					
Depth					
Moisture Standard.					
Density Standard					
Density Count					
Moisture Count					
Wet Density					
Dry Density					
% Moisture					
% Compaction					
Pass / fail					
Inspector					
Lab Tech					
Lab report #					

Consec. #					
Date					
Test #					
Chain					
Sta.					
CL Offset					
Lift #					
Material Type					
Proctor #					
Da					
Optimum Moisture					
Plasticy Index					
Liquid Limit					
Density Req.					
Depth					
Moisture Standard.					
Density Standard					
Density Count					
Moisture Count					
Wet Density					
Dry Density					
% Moisture					
% Compaction					
Pass / fail					
Inspector					
Lab Tech					
Lab report #					

Consec. #						
Date						
Test #						
Chain						
Sta.						
CL Offset						
Lift #						
Material Type						
Proctor #						
Da						
Optimum Moisture						
Plasticy Index						
Liquid Limit						
Density Req.						
Depth						
Moisture Standard.						
Density Standard						
Density Count						
Moisture Count						
Wet Density						
Dry Density						
% Moisture						
% Compaction						
Pass / fail						
Inspector						
Lab Tech						
Lab report #						

Consec. #						
Date						
Test #						
Chain						
Sta.						
CL Offset						
Lift #						
Material Type						
Proctor #						
Da						
Optimum Moisture						
Plasticy Index						
Liquid Limit						
Density Req.						
Depth						
Moisture Standard.						
Density Standard						
Density Count						
Moisture Count						
Wet Density						
Dry Density						
% Moisture						
% Compaction						
Pass / fail						
Inspector						
Lab Tech						
Lab report #						

Consec. #						
Date						
Test #						
Chain						
Sta.						
CL Offset						
Lift #						
Material Type						
Proctor #						
Da						
Optimum Moisture						
Plasticy Index						
Liquid Limit						
Density Req.						
Depth						
Moisture Standard.						
Density Standard						
Density Count						
Moisture Count						
Wet Density						
Dry Density						
% Moisture						
% Compaction						
Pass / fail						
Inspector						
Lab Tech						
Lab report #						

Consec. #					
Date					
Test #					
Chain					
Sta.					
CL Offset					
Lift #					
Material Type					
Proctor #					
Da					
Optimum Moisture					
Plasticy Index					
Liquid Limit					
Density Req.					
Depth					
Moisture Standard.					
Density Standard					
Density Count					
Moisture Count					
Wet Density					
Dry Density					
% Moisture					
% Compaction					
Pass / fail					
Inspector					
Lab Tech					
Lab report #					

Consec. #					
Date					
Test #					
Chain					
Sta.					
CL Offset					
Lift #					
Material Type					
Proctor #					
Da					
Optimum Moisture					
Plasticy Index					
Liquid Limit					
Density Req.					
Depth					
Moisture Standard.					
Density Standard					
Density Count					
Moisture Count					
Wet Density					
Dry Density					
% Moisture					
% Compaction					
Pass / fail					
Inspector					
Lab Tech					
Lab report #					

Consec. #					
Date					
Test #					
Chain					
Sta.					
CL Offset					
Lift #					
Material Type					
Proctor #					
Da					
Optimum Moisture					
Plasticy Index					
Liquid Limit					
Density Req.					
Depth					
Moisture Standard.					
Density Standard					
Density Count					
Moisture Count					
Wet Density					
Dry Density					
% Moisture					
% Compaction					
Pass / fail					
Inspector					
Lab Tech					
Lab report #					

Consec. #						
Date						
Test #						
Chain						
Sta.						
CL Offset						
Lift #						
Material Type						
Proctor #						
Da						
Optimum Moisture						
Plasticy Index						
Liquid Limit						
Density Req.						
Depth						
Moisture Standard.						
Density Standard						
Density Count						
Moisture Count						
Wet Density						
Dry Density						
% Moisture						
% Compaction						
Pass / fail						
Inspector						
Lab Tech						
Lab report #						

SAMPLE 2
LOST DAYS LOGS

LOG FOR LOST DAYS

DATE			DUE TO		RAIN		
MM	DD	YY	RAIN/MUD	OTHER	INCHES	FROM	TO
02	02	12		1	Freezing Temp.		
02	03	12		1	"		
02	04	12		1	"		
02	06	12		1	"		
02	07	12		1	"		
02	09	12		1	"		
02	10	12		1.	"		
02	11	12		1.	"		
02	14	12	1	.	0.50	14:00	15:30
02	15	12	1		0.35	12:00	18:00
02	20	12	1		1	18:00	20:00
02	21	12	1		0.50	10:00	13:00
02	22	12	1		Mud.	300+00	325+00
02	23	12	1		Mud	300+00	325+00
03	20	12	1		1	0:00	06:00
03	21	12	1		Mud	200+20	200+90
03	22	12	1		Mud	210+15	230+00
04	17	12	1		0.8	16:00	20:00
04	20	12		1			
Subtotal			10	9			

SECTION 2
LOST DAYS LOGS

DATE			DUE TO		RAIN		
MM	DD	YY	RAIN/MUD	OTHER	INCHES	FROM	TO
	Subtotal						

LOST DAYS (continued)

DATE			DUE TO		RAIN		
MM	DD	YY	RAIN/MUD	OTHER	INCHES	FROM	TO
	Subtotal						

Lost Days (continued)

DATE			DUE TO		RAIN		
MM	DD	YY	RAIN/MUD	OTHER	INCHES	FROM	TO
	Subtotal						

Lesson 1
BASIC SOIL PROPERTIES

The *plasticity index* of a soil is an indication of the clay content and the soil's moisture-retaining capabilities. A large plasticity index indicates high clay content and an increased ability to retain water, resulting in the tendency for the soil to swell when wet and shrink when dry. As a general rule, soils with plasticity indexes greater than 15 percent are "troublemakers." These soils require the addtion of a stabilization agent, such as lime, when used in subgrades. Using lime can substantially increase the stability, impermeability, and load-bearing capacity of the subgrade.

Please note that *clay* is a term that has been used to define both mineral type and particle size.

The plasticity index and liquid limits for any given soil are directly related to the in-place density obtained while compacting the soil.

The plasticity index of a given soil is the difference between the soil's plastic limit and liquid limit.

Plasticity Index (PI)% = Liquid Limit
(LL)% – Plastic Limit (PL)%

The *liquid limit* is the percentage of moisture content at which a soil changes, with decreasing wetness from the liquid to the plastic consistency or with increasing wetness from the plastic to the liquid consistency.

If the liquid limit of the portion of the sample passing the sieve no. 40 is less than 50 percent, the sample is considered to have a low compressibility. If the liquid limit is 50 percent or greater, the soil is considered to be high compressible

Compressibility is the degree to which a soil mass decreases in volume when supporting a load. Compressibility is lowest in coarse-grained soils where particles are in contact with each other. It increases as the proportion of small particles increases and becomes highest in fine-grained soils that contain organic matter. Following are examples of compressibility:

- Gravels and sands are practically incompressible. If a moist mass of these materials is subjected to compression, there is no significant change in their volume.
- Clays are compressible. If a moist mass of clay is subjected to compression, moisture and air may be expelled, resulting in a volume reduction that is not immediately recovered when the load is removed.

The *plastic limit* is the percentage of moisture content at which a soil changes with decreasing wetness from the plastic to the semisolid consistency or with increasing wetness from the semisolid to the plastic consistency. It is also the moisture content, expressed as a percentage of the dry sample weight at which the sample begins to crumble when rolled into a thread of 1/8-inch diameter.

The plastic limit is the lower limit of the plastic state. A small increase in moisture above the plastic limit will destroy the cohesion of the soil.

The plastic limit is a good indicator of the optimum moisture content, typically within 2 or 3 percent to 4 percent for material with a plasticity index greater than 35 (PI> 35 percent).

The *moisture content* determines the extent to which the soil can be compacted. Water lubricates soil so that soil particles can slide into a more compacted condition. Water also fills voids that otherwise would contain air only. However, too much water will cause soil particles to float out of the voids, resulting in decreasing density. Therefore, too little or too much water makes it impossible to achieve proper compaction. The moisture content at which maximum soil density can be obtained with a given compaction effort is called the *optimum moisture content.*

You can perform a simple test to determine if soil contains the right amount of water for good compaction. (Uniform gravel or mostly sandy do not react well to this test.) Take a handful of the soil and squeeze it into the size and shape of a tennis ball, then drop the soil ball to the ground from a distance of one foot. At optimum moisture content, the ball will break apart into a small number of fairly uniform fragments. If the soil is too dry, the soil will not form a ball without water being added to it. If the soil is too wet, it will not break apart when dropped.

Lesson 2
MOISTURE-DENSITY CURVE ACCEPTANCE LIMITS

The following charts describe the criteria for density and moisture acceptance as described in item 132 of the 2004 Txdot specification.

ACCEPTANCE MOISTURE FOR MATERIAL WITH A PLASTICITY INDEX BETWEEN 15% AND 35% (15%< PI <35%)

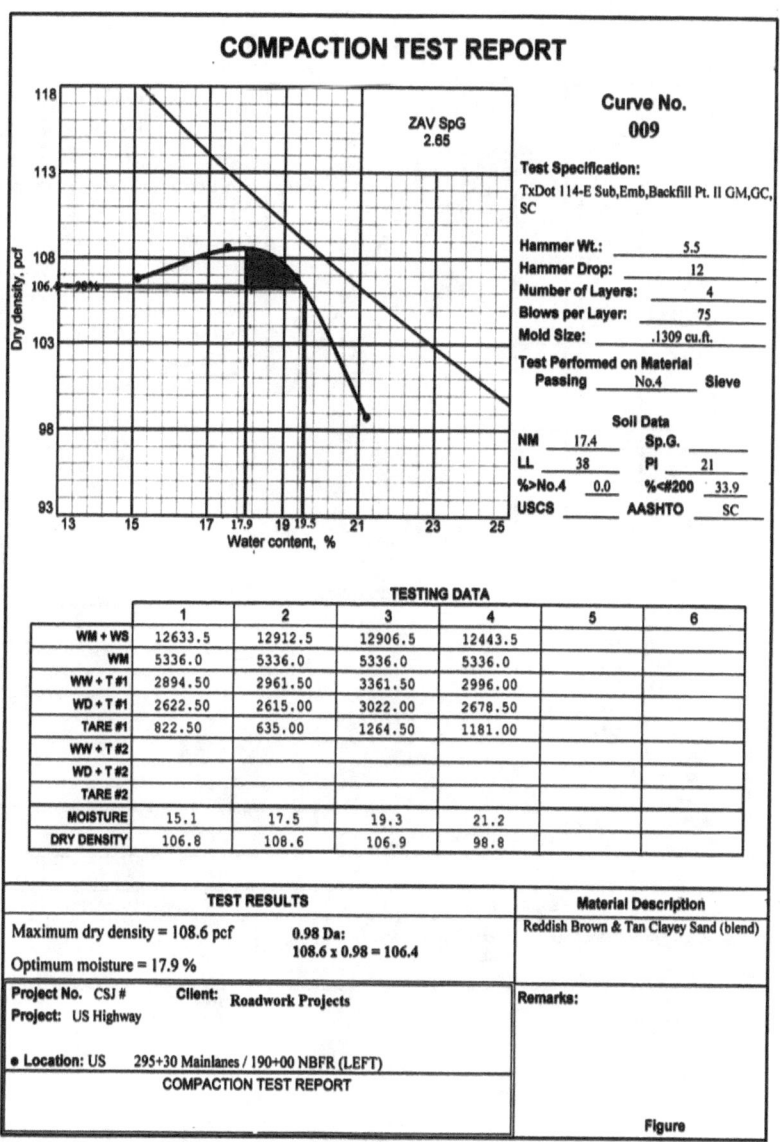

COMPACTION TEST REPORT

Curve No.
ZAV SpG 2.65
009

Test Specification:
TxDot 114-E Sub,Emb,Backfill Pt. II GM,GC, SC

Hammer Wt.:	5.5
Hammer Drop:	12
Number of Layers:	4
Blows per Layer:	75
Mold Size:	.1309 cu.ft.

Test Performed on Material
Passing ___No.4___ Sieve

Soil Data

NM	17.4	Sp.G.	
LL	38	PI	21
%>No.4	0.0	%<#200	33.9
USCS		AASHTO	SC

(Graph: Dry density, pcf vs Water content, % — y-axis 93 to 118, x-axis 13 to 25; marked points 17.9, 19.5)

TESTING DATA

	1	2	3	4	5	6
WM + WS	12633.5	12912.5	12906.5	12443.5		
WM	5336.0	5336.0	5336.0	5336.0		
WW + T #1	2894.50	2961.50	3361.50	2996.00		
WD + T #1	2622.50	2615.00	3022.00	2678.50		
TARE #1	822.50	635.00	1264.50	1181.00		
WW + T #2						
WD + T #2						
TARE #2						
MOISTURE	15.1	17.5	19.3	21.2		
DRY DENSITY	106.8	108.6	106.9	98.8		

TEST RESULTS	Material Description
Maximum dry density = 108.6 pcf **0.98 Da:** 108.6 x 0.98 = 106.4 Optimum moisture = 17.9 %	Reddish Brown & Tan Clayey Sand (blend)
Project No. CSJ # **Client:** Roadwork Projects **Project:** US Highway	**Remarks:**
● **Location:** US 295+30 Mainlanes / 190+00 NBFR (LEFT) COMPACTION TEST REPORT	
	Figure

ACCEPTANCE MOISTURE FOR MATERIAL WITH A
PLASTICITY INDEX GREATER THAN 35% (PI >35)

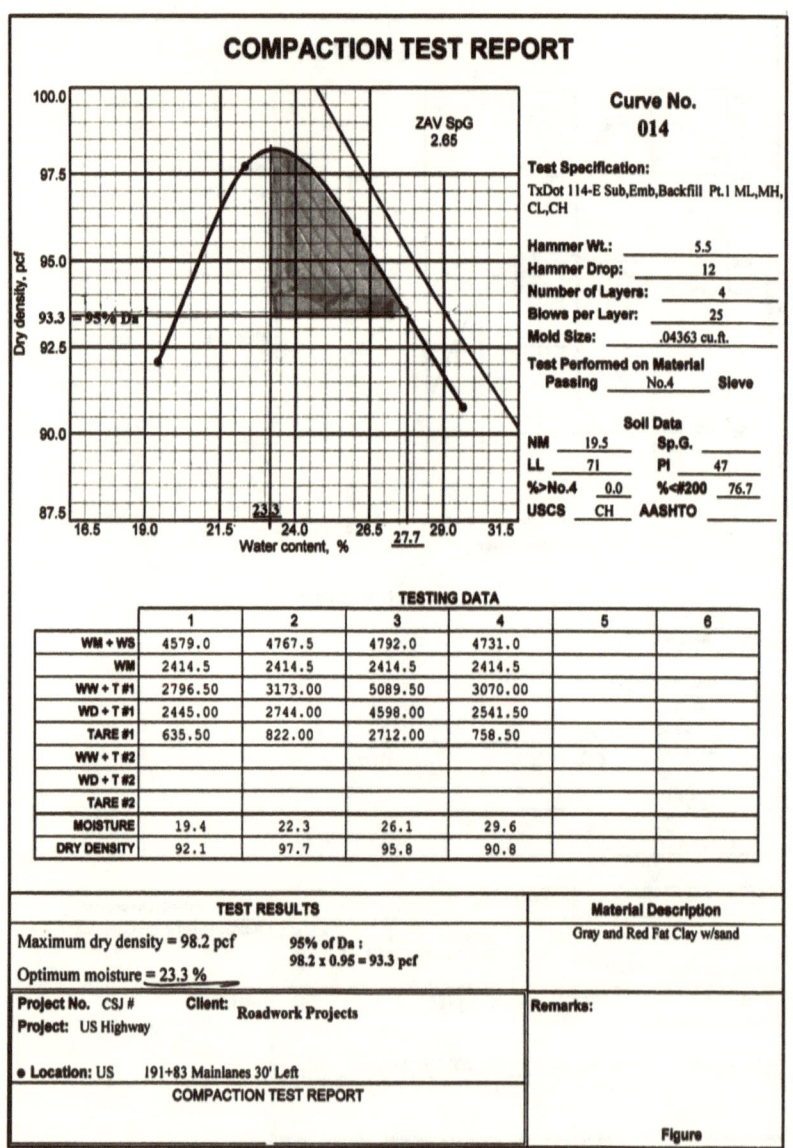

COMPACTION TEST REPORT

Curve No.
014

ZAV SpG
2.65

Test Specification:
TxDot 114-E Sub,Emb,Backfill Pt.1 ML,MH,
CL,CH

Hammer Wt.:	5.5
Hammer Drop:	12
Number of Layers:	4
Blows per Layer:	25
Mold Size:	.04363 cu.ft.

Test Performed on Material
Passing _____ No.4 _____ Sieve

Soil Data

NM	19.5	Sp.G.	
LL	71	PI	47
%>No.4	0.0	%<#200	76.7
USCS	CH	AASHTO	

TESTING DATA

	1	2	3	4	5	6
WM + WS	4579.0	4767.5	4792.0	4731.0		
WM	2414.5	2414.5	2414.5	2414.5		
WW + T #1	2796.50	3173.00	5089.50	3070.00		
WD + T #1	2445.00	2744.00	4598.00	2541.50		
TARE #1	635.50	822.00	2712.00	758.50		
WW + T #2						
WD + T #2						
TARE #2						
MOISTURE	19.4	22.3	26.1	29.6		
DRY DENSITY	92.1	97.7	95.8	90.8		

TEST RESULTS	Material Description
Maximum dry density = 98.2 pcf 95% of Da : 98.2 x 0.95 = 93.3 pcf Optimum moisture = 23.3 %	Gray and Red Fat Clay w/sand
Project No. CSJ # Client: Roadwork Projects Project: US Highway • Location: US 191+83 Mainlanes 30' Left COMPACTION TEST REPORT	Remarks: Figure

ACCEPTANCE MOISTURE FOR MATERIAL WITH A
PLASTICITY INDEX LESS THAN 15% (PI < 15%)

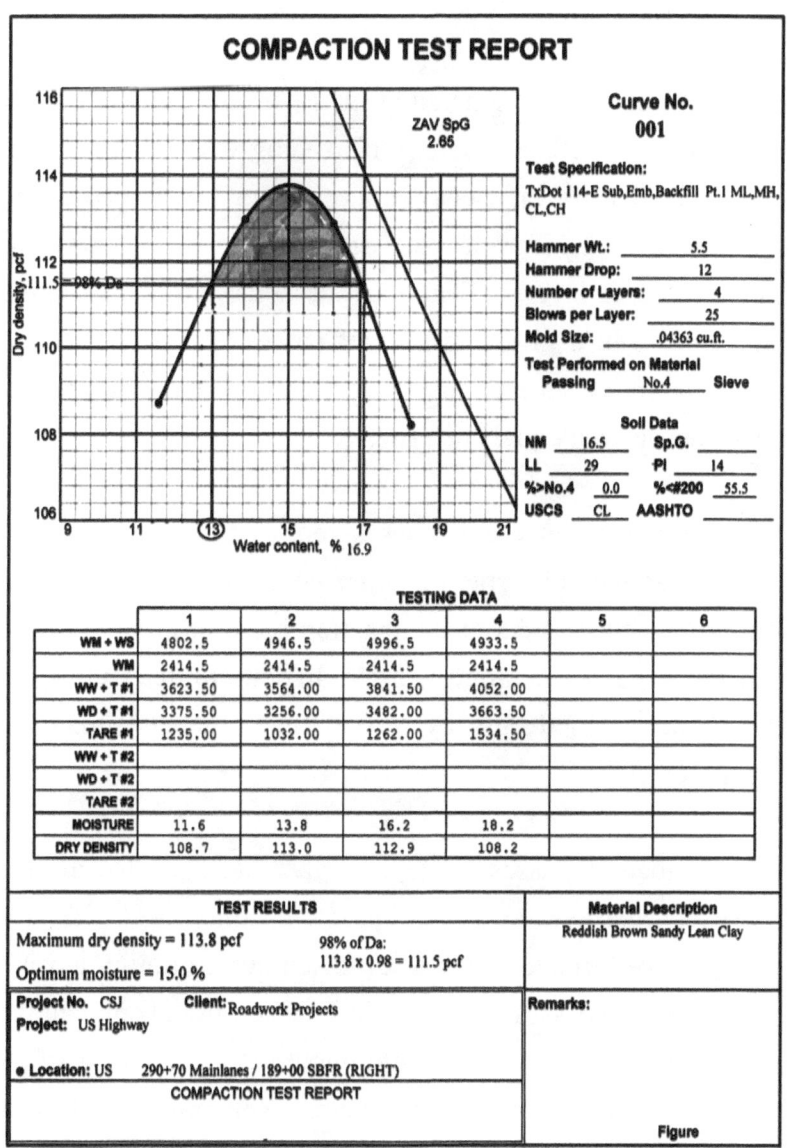

COMPACTION TEST REPORT

Curve No.

001

ZAV SpG 2.65

Test Specification:
TxDot 114-E Sub,Emb,Backfill Pt.1 ML,MH, CL,CH

Hammer Wt.:	5.5
Hammer Drop:	12
Number of Layers:	4
Blows per Layer:	25
Mold Size:	.04363 cu.ft.

Test Performed on Material
Passing ___No.4___ Sieve

Soil Data
NM	16.5	Sp.G.	
LL	29	PI	14
%>No.4	0.0	%<#200	55.5
USCS	CL	AASHTO	

Water content, % 16.9

TESTING DATA

	1	2	3	4	5	6
WM + WS	4802.5	4946.5	4996.5	4933.5		
WM	2414.5	2414.5	2414.5	2414.5		
WW + T #1	3623.50	3564.00	3841.50	4052.00		
WD + T #1	3375.50	3256.00	3482.00	3663.50		
TARE #1	1235.00	1032.00	1262.00	1534.50		
WW + T #2						
WD + T #2						
TARE #2						
MOISTURE	11.6	13.8	16.2	18.2		
DRY DENSITY	108.7	113.0	112.9	108.2		

TEST RESULTS

Maximum dry density = 113.8 pcf

Optimum moisture = 15.0 %

98% of Da:
113.8 x 0.98 = 111.5 pcf

Material Description

Reddish Brown Sandy Lean Clay

Project No. CSJ   Client: Roadwork Projects
Project: US Highway

Remarks:

● Location: US   290+70 Mainlanes / 189+00 SBFR (RIGHT)

COMPACTION TEST REPORT

Figure

Lesson 3
COMPACTION FUNDAMENTALS

The depth of each compacted layer is perhaps the single most important controllable factor influencing density. To obtain maximum density efficiency, material should be spread and compacted in layers not exceeding a depth of twenty-four inches. Deeper layers will reduce the density that a machine can develop in a given number of passes. If it is not stated otherwise in the specs, consider that each time the equipment rolls over an area equals one pass.

The number of machine passes required to compact soil depends on the soil type, moisture content, machine weight, and types and degrees of compaction required; therefore the number of passes required is often difficult to determine and can be determined only by testing the compacted material density onsite.

The number of passes made over the material also affects density. Regardless of what type of machine used, the unit should make three to four passes to achieve optimum density. More than five passes will result in little additional compactive effort; the added expense of additional passes is not justified by the incremental increase in density. Remember, each pass means more use of fuel, oil, and operator time as well as wear and tear on the equipment.

To maximize compaction, a compactor should be operated on as flat a slope as possible. This is because of the weight of the soil compactor is more efficiently used and concentrated when working on a flat surface.

Soil compactors should never operate on a slope steeper than 4:1. Soil continues to compact until the internal resistance within the soil reaches equilibrium with the compaction pressure.

By using special compaction equipment, such as compactor wheels or tamper-plate attachments in excavators, or heavy compaction equipment in the field, the maximun in-place soil density is often greater than 100 percent of the maximum dry density obtained in the testing laboratory under controlled conditions. However, avoid overcompacting the soil, since this wastes time and results in unnesessary equipment wear and fuel consumption. Also, the soil will begin to shift sideways.

Field densities greater than 103% may indicate that the used Proctor is not representative of the material being tested. Also, some specifications states that no field density test shall be outside the limits by more than 3 pounds per cubic feet. Therefore, is important to understand the properties of the material and to know the limits for acceptance indicated on the specifications

LESSON 4
SOIL SWELL AND SHRINK

Soil exists physically in three fundamental states:

1. Loose (in a stockpile)
2. In the bank (natural, undisturbed condition)
3. Compacted

Soil swells when excavated and shrinks under pressure, and volume corrections (swell and shrinkage factors) must often be used when estimating quantities or production rates.

All earth material that excavation contractors handle can exist in at least three different volumes of measure: bank volumes, trucked volume, and compacted volume. So if you need to build a pad for a building and the volume is 1,000 CCY (compacted cubic yards), you will need to know how many BCY (bank cubic yards) you will have to extract and how many TCY (trucked cubic yards) you will need to haul, because those numbers are not the same. One of the most common and serious estimate mistakes is to misadjust or not adjust the measured quantities of cut and/or fill for the necessary swell and shrink of the material being estimated.

To correctly apply the swell and shrink adjustments, the estimator must understand the basic properties of materials. All earthen materials, from loam to granite, contain the following components:

- solid particles
- voids or air pockets
- moisture

The density of materials refers to the ratio of solid particles to voids. In other words, a cubic yard of material at 90 percent true density would consist of 90 percent solid particles and 10 percent voids by volume. Moisture content is a measure of the amount or percentage of the void space that contains water.

Earthmoving operations always change the density of the material and usually have some effect on the moisture content. When we cut, dig, or otherwise disturb materials, we reduce the density and increase the amount of void space, which is is known as "swell." When we compact materials, we reduce the amount of void space, which is known as "shrink." The figures in the following chart depict this change proportionally for a cubic yard of damp clay.

For example. how much damp clay material needs to be borrowed to fill a 12,000 CY embankment? The clay is to be compacted to 98 percent density.

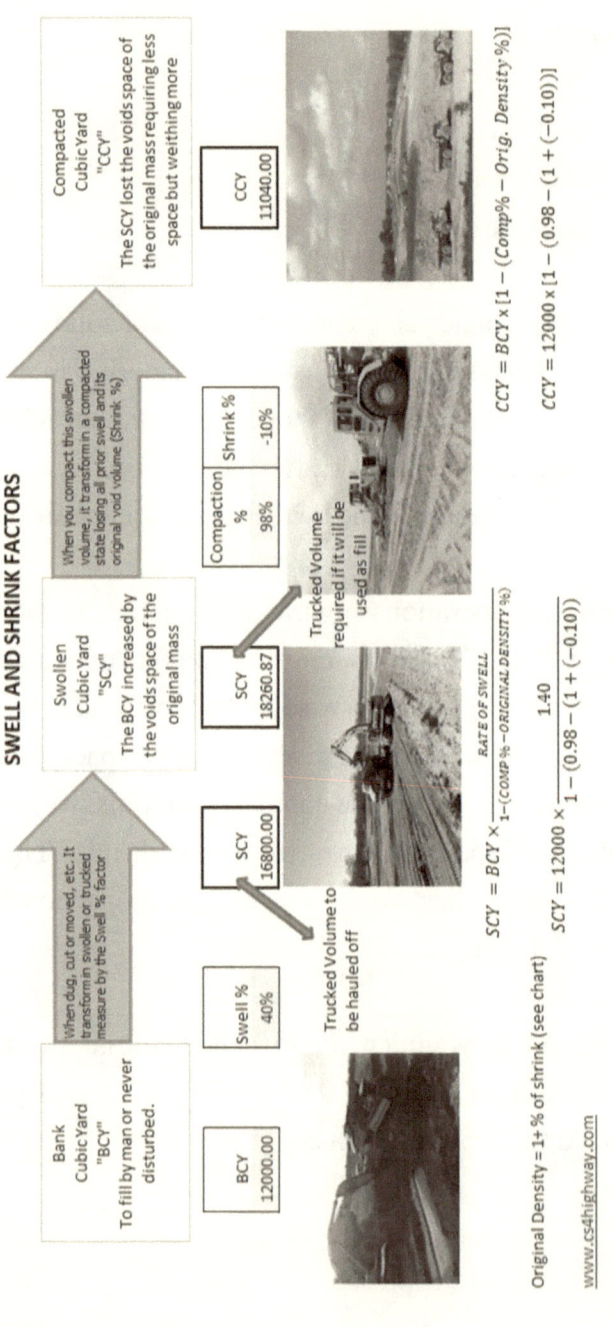

SWELL AND SHRINK FACTORS

Bank Cubic Yard "BCY"
To fill by man or never disturbed.

When dug, cut or moved, etc. It transform in swollen or trucked measure by the Swell % factor

Swollen Cubic Yard "SCY"
The BCY increased by the voids space of the original mass

When you compact this swollen volume, it transform in a compacted state losing all prior swell and its original void volume (Shrink %)

Compacted Cubic Yard "CCY"
The SCY lost the voids space of the original mass requiring less space but weithing more

BCY	Swell %	SCY	SCY	Compaction %	Shrink %	CCY
12000.00	40%	16800.00	18260.87	98%	-10%	11040.00

Trucked Volume to be hauled off

Trucked Volume required if it will be used as fill

$$SCY = BCY \times \frac{RATE\ OF\ SWELL}{1-(COMP\ \% - ORIGINAL\ DENSITY\ \%)}$$

$$SCY = 12000 \times \frac{1.40}{1-(0.98-(1+(-0.10))}$$

$$CCY = BCY \times [1-((Comp\% - Orig.\ Density\ \%)]$$

$$CCY = 12000 \times [1-((0.98-(1+(-0.10))]$$

Original Density = 1+ % of shrink (see chart)

www.cs4highway.com

SWELL %, SHRINK%, AND WEIGHTS
OF COMMON MATERIALS

Type of material	Weight in pounds per bank CY (natural state)	% swell when excavated or loaded	Percent of shrink (-) or swell (Natural State 100%comp)
Earth			
Clay			
Dry	2900	35	-10
Damp	3200	40	-10
Gravel			
Dry	3000	15	-7
Pit Run	3400	8	-4
Wet	3500	5	-3
Mud	2600	20	-15
Sand			
Dry	2900	10	-10
Wet	3100	5	-10
Silt	3000	35	-20
Topsoil	2400	55	-25
Rock			
Basalt	5000	60	35
Decomposed rock			
75% Rock,25% Earth	4100	40	30
50% Rock, 50% Earth	3800	27	0
25% Rock, 75% Earth	3400	25	-5

Dolomite	4900	65	30
Feldspar	4400	65	30
Gneiss	4500	65	30
Granite	4500	70	35
Limestone	4300	60	30
Quartzite	4500	65	35
Sandstone	4400	60	25
Taconite, iron ore	5500	60	30
Trap rock	4700	65	30
Miscellaneous			
Cinders	1300	30	-10
Ice	1600	70	NA
Pavement removal			
Asphalt	3200	50	20
Concrete	3900	65	30
Peat	1200	30	-25
RipRap (avg.)	4500	70	40
Snow			
Dry	250	0	NA
Wet	850	0	NA

Lesson 5
CLASSIFICATION OF SOILS FOR ENGINEERING PURPOSES

The widely used Unified Soil Classification System (USCS) gives each soil a two-letter designation, where the first letter describes the major soil constituents and the second letter describes the soil's gradation, or plasticity. The five major soil constituentes are abbreviatiated as gravel, G; sand, S; silt, M; clay, C; and organic matter, O. Abbrevations used to describe gradation and plasticity are well-graded, W; poorly graded, P; low plasticity, L; and high plasticity, H.

Clay is a fine-grained soil that can be made to exhibit plasticity (puttylike properties) within a range of water contents and that exhibits considerable strength when dried by the air. There is a simple test that you can perform to determine if a soil has a high clay content. Roll a moist sample of soil into a ball approximately one inch in diameter and throw it against a wall. If the soil has a high clay content, the ball will stick to the wall.

Silt is soil passing a No. 200 sieve that is nonplastic or very slightly plastic and that exhibits little or no strength when air-dried.

Organic matter is partially decomposed vegetable matter that will continue to decompose with time and leave voids; therefore, it should be removed from soil that is to be compacted.

Organic clay is a soil that would be classified as a clay except that its LL after oven drying (dry sample preparation) is less then 75 percent of its LL before oven drying (wet sample preparation).

Organic silt is a soil that would be classified as silt except that its LL after oven drying (dry sample preparation) is less then 75 percent of its LL before oven drying (wet sample preparation).

Peat is a soil composed of vegetable tissue in varios stages of decomposition, usually with dark brown to black color, a spongy consistency, and a texture ranging from resembling fibers to having no definite or clear shape or form.

Gravel consists of unconsolidated or loose detrital sediment (aggregate resulting from natural desintegration and abrasion of rock) with particle sizes passing through a three-inch sieve and retained in a No. 10 sieve.

Sand consists of fine aggregate particles that are retained in the No. 200 sieve, either as natural sand resulting from natural disintegration and abrasion of rock, or as manufactured sand, which is produced by crushing rock, gravel, slag, etc.

Soil grading is used to determine the distribution of grain size in the soil. The sieves are linked (racked) as shown in the picture below.

STANDARD U.S SIEVES AND PANS

Milimeters	Inches
75	3
63	2 1/2
50	2
37.5	1 1/2
25	1
19	3/4
12.5	1/2
9.5	3/8
4.75	No. 4
2.36	No. 8
1.18	No. 16
600μm	No. 30
300μm	No. 50
150μm	No. 100
75μm	No. 200

Symbol	Description	Stability as Construction Material
Coarse-Grained Soils (Less than 50% Passes through No. 200 Sieve)		
GW	Well-graded gravel	Excellent
SW	Well-graded sand	Excellent
GP	Poorly-graded gravel	Excellent to good
SP	Poorly-graded sand	Good
GM	Silty gravel	Good
SM	Silty sand	Fair
GC	Clayey gravel	Good
SC	Clayey sand	Good
Fine-Grained Soils (50% or More Passes through No. 200 Sieve)		
ML	Low-plasticity Silt	Fair
CL	Low-plasticity Clay	Good to fair
OL	Low-plasticity Organic	Fair
MH	High-plasticity silt	Poor
CH	High-plasticity clay	Poor
OH	High-plasticity organic	Poor
PT	Peat or organic	Unsuitable

Lesson 6
HOW TO BUILD AN EMBANKMENT EFFECTIVELY

A young leadman was struggling with an excavation and embankment operation that involved clay material. "What do you think?" he asked. Well, clay material is never easy to work with, but I had seen in the past how other contractors have dealt with the entire process and effectively succeed on building the road. So I gladly described the process to him in the following way, hoping he would find a solution to his operation problems:

First, you need to know the properties of the material that you are dealing with. Your quality-control team should be able to provide you with this information. Remember that you need to notify the inspection representatives before opening a material source to allow them to perform quality-assurance testing. Provide advance notice, since it takes some time to get the results of the sampling and testing.

Once you have selected the material source and determined how many cubic yards you will need to haul into the fill area, it is most effective for compactors' production to select at least two or three areas of embankment that can be built simultaneously. Begin working the subgrade. For good compaction, the subgrade needs to be scarified and loosened. This applies to all unpaved surface areas, except rock, to a depth of at least six inches.

Haul in and spread the first lift of material over the first area. Move the material dumped in piles or windrows by blading or by similar methods, and incorporate it into uniform layers. Feather-edge or mix abutting layers of dissimilar material for at least one hundred feet to ensure there are no abrupt changes in the material. Break down clods or lumps of material, and mix embankment until a uniform material is attained.

At the same time, start preparing the second subgrade area. Remember to bench slopes before placing material, if required. Begin the placement of material at the toe of slopes. Do not place trees, stumps, roots, vegetation, or other objectionable material in the embankment. Simultaneously recompact scarified material with the placed embankment material. It is also very important to maintain drainage in excavated areas to avoid damage to the roadway section.

As the first lift of fill material is being spread over the second area, the first lift of fill in the first area can be watered and compacted. Apply water free of industrial wastes and other objectionable matter to achieve the uniform moisture content specified for compaction. Compact the layer to the required density.

Determine the maximum lift thickness based on the ability of the compacting operation and equipment to meet the required density. Do not exceed the layer depth specified, usually sixteen inches loose or twelve inches compacted. Maintain a level layer to ensure uniform compaction.

Then the compactor can compact the first lift in the second area as the second lift is spread over in the first area. Proceed in this manner until all fill is brought up to final grade.

Construct embankments to the grade and sections shown on the plans. Construct the embankment in layers approximately parallel to the finished grade for the full width of the individual roadway cross sections. Ensure that each section of the embankment conforms to the detailed sections or slopes. This is achieved by installing color-coded stakes and ribbons in the cross section at every fifty feet. For example, use green stakes for the center line of the ditch, pink for the edge of the shoulder, and orange for the edge of the right of way. These stakes represent the embankment or excavation limits. To control the grade elevation, you may use blue ribbons to show the fill or cut on grade; red ribbons to show 1-foot cut or 1-foot fill from the ribbon; white ribbon to mean cut more than 2 feet; and green ribbon to mean fill more than 2 feet. The stakes should also include information about the required cut or fill measure.

This procedure allows the earth-moving crews to establish cycles and a methodology for working the dirt. An embankment operation cycle includes extracting the material, hauling the material to the fill area, spreading the material, processing the material to reach compaction, and preparing new subgrade for the next lift. The idea is to be able to create efficient cycles during each process.

Additional planning is required to extract the material at the pit properly, maintain good conditions on the haul roads, and have the right amount and type of equipment at all processes. You will have to have enough compaction equipment to keep up with the rate at which fill material is hauled in.

When compaction and moisture requirements on each lift are met, that is the go-ahead that contractors need to continue placing layers of dirt. Maintaining the finished section, density, moisture and grade until the project is accepted is the biggest challenge earthwork contractors have to overcome.

The density requirements that have to be met according to the Texas Department of Transportation are as follows:

Description	Density	Moisture Content
	Tex-115-E	
PI ≤ 15	≥ 98% D_a	
15 < PI ≤ 35	≥ 98% D_a and ≤ 102% D_a	≥ $W_{opt.}$
PI > 35	≥ 95% D_a and ≤ 100% D_a	≥ $W_{opt.}$

Once all requirements in the table above are met, you need to maintain the density and moisture content. For soils with a PI greater than 15, maintain the moisture content no lower than

4 percentage points below optimum. Rework the material to obtain the specified compaction when it loses the required stability, density, moisture, or finish. This is where the fun begins, because sometimes we have to alter the compaction methods and procedures on subsequent work to obtain the density and moisture requirements.

But how do you adjust moisture content? If the soil is too wet, it can be loosened and aerated with rippers or disk harrow, and then rerolled. If the soil failed the density test because it was too wet, it can sometimes be left in place to dry, rerolled, and tested again. In the meantime, you can be working on the second area, described earlier. This is why you need to plan several areas to work at the same time.

If, however, the test fails because it is too dry, you must rip it or disk it, add water, and reroll it with a tamping foot or sheepsfoot compactor that has the ability to work the water into the soil. Try hard to compact a lift as soon as possible after water has been added. If you take too long to do it, the upper portion of the lift will evaporate more than the bottom portion. Compacting a dry top layer over a wetter bottom layer will cause the top layer to crumble, leaving a scaly surface.

Tables

Table 1: Size Of Trench

The end area of the trench depends on the size of pipe installed and the bank slope required. According to the American Concrete Pipe Association, the bottom width required (A in the figure) is determined by the formula:

A = 1.25 Bc + 1, where Bc is the outside diameter of the pipe in feet

Trench Bottom Widths

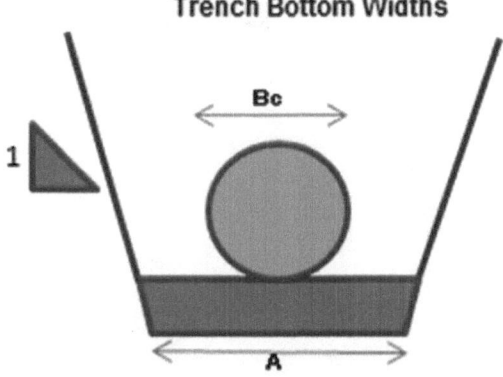

Pipe Diameter		Trench Width		Pipe Diameter		Trench Width	
mm	in	m	ft	mm	in	m	ft
102	4	0.49	1.6	1524	60	2.59	8.5
152	6	0.55	1.8	1676	66	2.80	9.2
203	8	0.61	2.0	1829	72	3.05	10.0
254	10	0.70	2.3	1981	78	3.26	10.7
305	12	0.76	2.5	2134	84	3.47	11.4
381	15	0.91	3.0	2286	90	3.69	12.1
457	18	1.03	3.4	2438	96	3.93	12.9
533	21	1.16	3.8	2591	102	4.15	13.6
610	24	1.25	4.1	2743	108	4.36	14.3
686	27	1.37	4.5	2896	114	4.54	14.9
838	33	1.58	5.2	3048	120	4.75	15.6
914	36	1.70	5.6	3200	126	4.99	16.4
1067	42	1.92	6.3	3353	132	5.21	17.1
1219	48	2.13	7.0	3505	138	5.43	17.8
1372	54	2.38	7.8	3658	194	5.64	18.5

TABLE 2: ANGLE OF REPOSE FOR COMMON SOIL TYPES

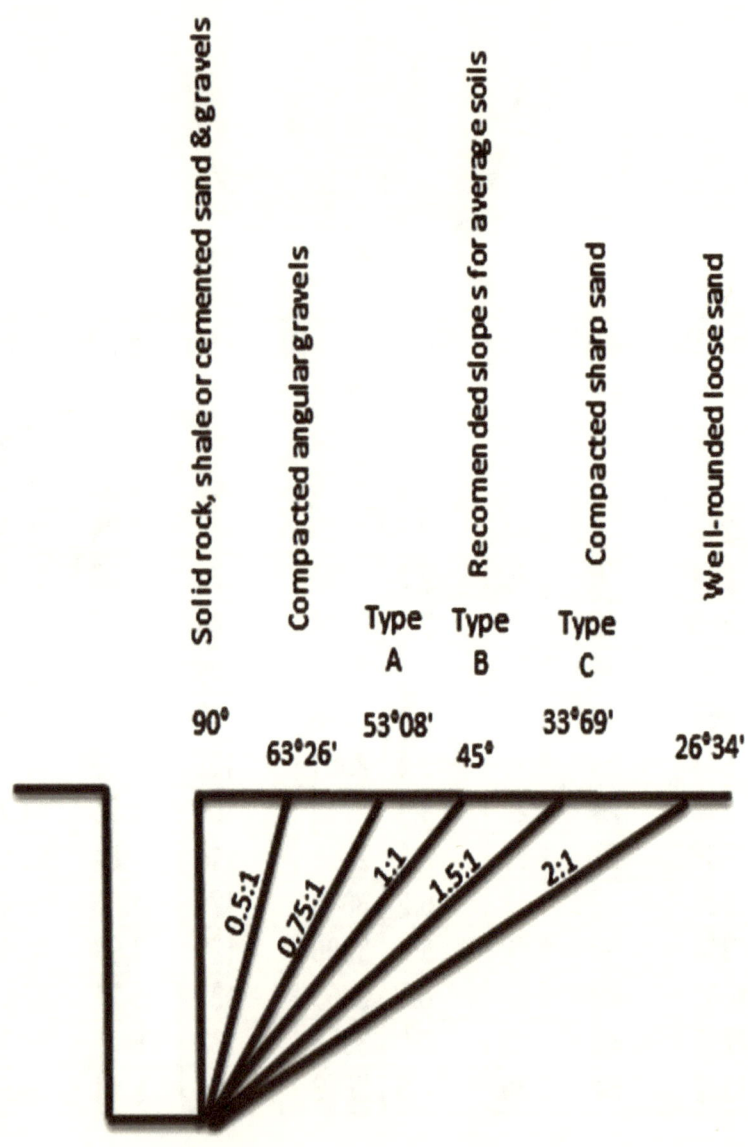

TABLE 3: COMMON SLOPES VALUES

SLOPE X:Y	Horizontal X	Vertical Y	Percentage %	Degrees α
1:1	1	1	100%	45.00
1.5:1	1.5	1	66.67%	33.69
2:1	2	1	50.00%	26.57
2.5:1	2.5	1	40.00%	21.80
3:1	3	1	33.33%	18.43
3.5:1	3.5	1	28.57%	15.95
4:1	4	1	25.00%	14.04
4.5:1	4.5	1	22.22%	12.53
6:1	6	1	16.67%	9.46
6.5:1	6.5	1	15.38%	8.75

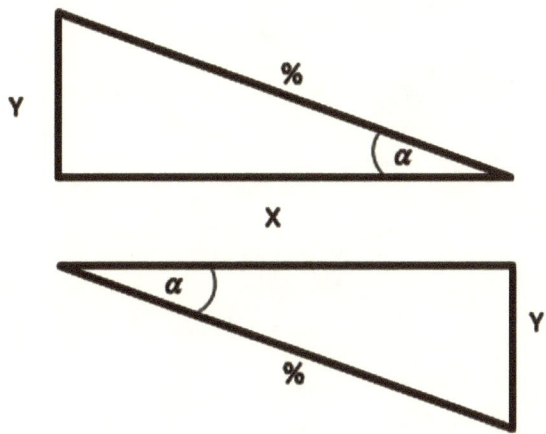

About The Author

Alberto Munguía Mireles is known in the projects as the Mexican Engineer "El Ingeniero". He is the most sought after breaking down construction processes for monitoring and controlling purposes. He is famous for his high standards of behavior, commitment to quality, dedication to personal fulfillment and help other to attain it. He stands against lack of commitment to planning, poor quality and acceptance of incompetence

Alberto Munguía Mireles is a Civil Engineer by trade and Project Management Professional by specialization. He has more than twenty four years experience providing engineering services during the construction and inspection of infrastructure projects. Since 2010 he has been dedicated to aggregate civil engineering, construction management, quality control and quality assurance methods used by experts during the construction and inspection of highways in Texas. He is well know for his expertise in monitoring and controlling construction processes and his ability to mitigate negative risk

Mr. Munguía earned his bachelor degree in the Ibero-American University. The Ibero is a prestigious Mexican institution of higher education sponsored by the Society of Jesuits. He holds international certificates granted by the American Concrete Institute and the Project Management Institute. He has also earned certificates issued by the Asphalt Hot Mix Association, the Construction Estimating Institute and the Bureau Veritas Mexicana S.A de C.V

Drawing on the Ibero legacy of improving ourselves and his integral and human formation, Mr. Munguía is committed to provide training to contractors, subcontractors, young engineers and students about subjects related to improve the overall quality of roadwork construction by sharing knowledge among them, providing guidance to lesser experienced personnel and teaching them how to use available tools and techniques used by roadwork experts